Third Edition

APPLIED SOCIAL RESEARCH METHODS SERIES

Series Editors

LEONARD BICKMAN, Peabody College, Vanderbilt University, Nashville
DEBRA J. ROG, Westat

1. **SURVEY RESEARCH METHODS (Fourth Edition)** by FLOYD J. FOWLER, Jr.
2. **SYNTHESIZING RESEARCH (Third Edition)** by HARRIS COOPER
3. **METHODS FOR POLICY RESEARCH** by ANN MAJCHRZAK
4. **SECONDARY RESEARCH (Second Edition)** by DAVID W. STEWART and MICHAEL A. KAMINS
5. **CASE STUDY RESEARCH (Fourth Edition)** by ROBERT K. YIN
6. **META-ANALYTIC PROCEDURES FOR SOCIAL RESEARCH (Revised Edition)** by ROBERT ROSENTHAL
7. **TELEPHONE SURVEY METHODS (Second Edition)** by PAUL J. LAVRAKAS
8. **DIAGNOSING ORGANIZATIONS (Second Edition)** by MICHAEL I. HARRISON
9. **GROUP TECHNIQUES FOR IDEA BUILDING (Second Edition)** by CARL M. MOORE
10. **NEED ANALYSIS** by JACK McKILLIP
11. **LINKING AUDITING AND META EVALUATION** by THOMAS A. SCHWANDT and EDWARD S. HALPERN
12. **ETHICS AND VALUES IN APPLIED SOCIAL RESEARCH** by ALLAN J. KIMMEL
13. **ON TIME AND METHOD** by JANICE R. KELLY and JOSEPH E. McGRATH
14. **RESEARCH IN HEALTH CARE SETTINGS** by KATHLEEN E. GRADY and BARBARA STRUDLER WALLSTON
15. **PARTICIPANT OBSERVATION** by DANNY L. JORGENSEN
16. **INTERPRETIVE INTERACTIONISM (Second Edition)** by NORMAN K. DENZIN
17. **ETHNOGRAPHY (Third Edition)** by DAVID M. FETTERMAN
18. **STANDARDIZED SURVEY INTERVIEWING** by FLOYD J. FOWLER, Jr., and THOMAS W. MANGIONE
19. **PRODUCTIVITY MEASUREMENT** by ROBERT O. BRINKERHOFF and DENNIS E. DRESSLER
20. **FOCUS GROUPS (Second Edition)** by DAVID W. STEWART, PREM N. SHAMDASANI, and DENNIS W. ROOK
21. **PRACTICAL SAMPLING** by GART T. HENRY
22. **DECISION RESEARCH** by JOHN S. CARROLL and ERIC J. JOHNSON
23. **RESEARCH WITH HISPANIC POPULATIONS** by GERARDO MARIN and BARBARA VANOSS MARIN
24. **INTERNAL EVALUATION** by ARNOLD J. LOVE
25. **COMPUTER SIMULATION APPLICATIONS** by MARCIA LYNN WHICKER and LEE SIGELMAN
26. **SCALE DEVELOPMENT (SECOND EDITION)** by ROBERT F. DeVELLIS
27. **STUDYING FAMILIES** by ANNE P. COPELAND and KATHLEEN M. WHITE
28. **EVENT HISTORY ANALYSIS** by KAZUO YAMAGUCHI
29. **RESEARCH IN EDUCATIONAL SETTINGS** by GEOFFREY MARUYAMA and STANLEY DENO
30. **RESEARCHING PERSONS WITH MENTAL ILLNESS** by ROSALIND J. DWORKIN
31. **PLANNING ETHICALLY RESPONSIBLE RESEARCH** by JOAN E. SIEBER
32. **APPLIED RESEARCH DESIGN** by TERRY E. HEDRICK, LEONARD BICKMAN, and DEBRA J. ROG
33. **DOING URBAN RESEARCH** by GREGORY D. ANDRANOVICH and GERRY RIPOSA
34. **APPLICATIONS OF CASE STUDY RESEARCH** by ROBERT K. YIN
35. **INTRODUCTION TO FACET THEORY** by SAMUEL SHYE and DOV ELIZUR with MICHAEL HOFFMAN
36. **GRAPHING DATA** by GARY T. HENRY
37. **RESEARCH METHODS IN SPECIAL EDUCATION** by DONNA M. MERTENS and JOHN A. McLAUGHLIN
38. **IMPROVING SURVEY QUESTIONS** by FLOYD J. FOWLER, Jr.
39. **DATA COLLECTION AND MANAGEMENT** by MAGDA STOUTHAMER-LOEBER and WELMOET BOK VAN KAMMEN
40. **MAIL SURVEYS** by THOMAS W. MANGIONE
41. **QUALITATIVE RESEARCH DESIGN** by JOSEPH A. MAXWELL
42. **ANALYZING COSTS, PROCEDURES, PROCESSES, AND OUTCOMES IN HUMAN SERVICES** by BRIAN T. YATES
43. **DOING LEGAL RESEARCH** by ROBERT A. MORRIS, BRUCE D. SALES, and DANIEL W. SHUMAN
44. **RANDOMIZED EXPERIMENTS FOR PLANNING AND EVALUATION** by ROBERT F. BORUCH
45. **MEASURING COMMUNITY INDICATORS** by PAUL J. GRUENEWALD, ANDREW J. TRENO, GAIL TAFF, and MICHAEL KLITZNER
46. **MIXED METHODOLOGY** by ABBAS TASHAKKORI and CHARLES TEDDLIE
47. **NARRATIVE RESEARCH** by AMIA LIEBLICH, RIVKA TUVAL-MASHIACH, and TAMAR ZILBER
48. **COMMUNICATING SOCIAL SCIENCE RESEARCH TO POLICY-MAKERS** by ROGER VAUGHAN and TERRY F. BUSS
49. **PRACTICAL META-ANALYSIS** by MARK W. LIPSEY and DAVID B. WILSON
50. **CONCEPT MAPPING FOR PLANNING AND EVALUATION** by MARY KANE and WILLIAM M. K. TROCHIM
51. **CONFIGURATIONAL COMPARATIVE METHODS** by BENOÎT RIHOUX and CHARLES C. RAGIN

Theory and Applications

Third Edition

Robert F. DeVellis
University of North Carolina, Chapel Hill

26 APPLIED SOCIAL RESEARCH METHODS SERIES
Edited by Leonard Bickman and Debra J. Rog

Los Angeles | London | New Delhi
Singapore | Washington DC

Los Angeles | London | New Delhi
Singapore | Washington DC

FOR INFORMATION:

SAGE Publications, Inc.
2455 Teller Road
Thousand Oaks, California 91320
E-mail: order@sagepub.com

SAGE Publications Ltd.
1 Oliver's Yard
55 City Road
London EC1Y 1SP
United Kingdom

SAGE Publications India Pvt. Ltd.
B 1/I 1 Mohan Cooperative Industrial Area
Mathura Road, New Delhi 110 044
India

SAGE Publications Asia-Pacific Pte. Ltd.
33 Pekin Street #02-01
Far East Square
Singapore 048763

Acquisitions Editor: Vicki Knight
Editorial Assistant: Kalie Koscielak
Production Editor: Astrid Virding
Copy Editor: Megan Speer
Typesetter: C&M Digitals (P) Ltd.
Proofreader: Scott Oney
Indexer: Ellen Slavitz
Cover Designers: Anupama Krishnan, Candice Harman
Marketing Manager: Helen Salmon
Permissions Editor: Adele Hutchinson

Printed in the United States of America

Library of Congress Cataloging-in-Publication Data

DeVellis, Robert F. Scale development : theory and applications/
Robert F. DeVellis.—3rd ed.

p. cm.—(Applied social research methods series; 26)
Includes bibliographical references and index.

ISBN 978-1-4129-8044-9 (pbk.)

1. Scaling (Social sciences). I. Title.

H61.27.D48 2012
300.72—dc22
2011002322

This book is printed on acid-free paper.

11 12 13 14 15 10 9 8 7 6 5 4 3 2

Contents

Preface

From the time I first contemplated writing a book on scale development, my goal has been to present a rather complex body of information in a way that helps readers understand the logic underlying the creation, use, and evaluation of measurement instruments and to develop a more intuitive feel for how scales work. I have sought to demystify measurement by relating it to more familiar experiences when possible and by emphasizing a conceptual rather than a strictly mathematical understanding. I have also endeavored to gather up and share the insights I have acquired over a career of work in this field—some bestowed on me by teachers and colleagues, others obtained only after long struggles with initially misunderstood concepts. The widespread adoption of this text in its earlier editions across a variety of disciplines has been gratifying. Former students have shared anecdotes of encountering copies of *Scale Development* in unexpected and somewhat surprising places, such as on the desk of a NASA scientist and in numerous faraway locales on several continents. I think what people have found useful is my emphasis on presenting the material in a way that is accessible to learners with relatively little prior background but still informative to those with considerably more previous measurement experience. That is again my intent in this edition. To that end, I have kept what continues to work well while updating the treatment of several topics, expanding the treatment of others, and providing newer and clearer examples of concepts when warranted.

Chapter 1, concerning the historical origins of measurement, has been expanded and now includes some additional historical perspectives on measurement. Notably, I have added a description of how early scientists, including Sir Isaac Newton, regarded inconsistencies in observations—what we would now call measurement error—as a moral failing to be hidden from colleagues. In this chapter and elsewhere, I acknowledge more directly the contributions of Lord and Novick's landmark 1968 volume, *Statistical Theories of Mental Test Scores,* which recently became more widely available after decades of scarcity.

In Chapter 2, an introduction to the latent variable, I also have added new material, including a consideration of the difference between aspects of the environment and respondents' perceptions of the environment—two distinct variables that are sometimes confused. I have also clarified the subtle distinction between tau-equivalent and *essentially* tau-equivalent models.

Chapter 3, on reliability, has been expanded to consider scale reliability in its broader context, as the degree to which an observed phenomenon reflects some consistent process. This broader approach considers the intraclass correlation coefficient (ICC) as a prototypic means of expressing the proportion of signal to signal-plus-noise and conceptually links the ICC to more specific methods for expressing reliability, including the computation of Cronbach's alpha and Cohen's kappa. Kappa crops up in the context of discussing inter-rater reliability. Although kappa is somewhat tangential to scale reliability in the narrowest sense, students have often failed to connect reliability as it relates to observers with reliability as it relates to items. Connecting those dots, I believe, enhances the understanding of reliability more generally. I also discuss some of the recent criticisms of Cronbach's alpha as a means of expressing reliability.

Chapter 4, on validity, benefits from added examples of how validity can be established. I also have included additional information on distinguishing conceptually between criterion and construct validity and between content and face validity. I have added a brief discussion of correlation attenuation and its implications for validity assessment.

Chapter 5 remains a list of steps one can take in developing a scale. I have expanded the discussion of item redundancy, differentiating more clearly between "good" and "bad" redundancy within a scale.

In Chapter 6, I have expanded the coverage of factor analysis and extensively revised several examples. New figures accompany those examples and, I believe, will further clarify concepts such as factor rotation. I also include a discussion of parallel analysis as a tool for determining how many factors to retain.

Chapter 7 has been substantially expanded to reflect the growing emphasis on item response theory (IRT) in the past few years. Although this chapter is still an overview rather than a detailed examination of IRT, I have now included a discussion of multiple response option IRT models and have added figures to illustrate how to interpret the graphical output that often accompanies the use of IRT. I have also extended the discussion of the role of classical test methods vis-à-vis IRT approaches and have summarized the findings of studies that have compared the two approaches directly.

Chapter 8 updates sources of information on existing measurement scales and includes some information on the Patient-Reported Outcomes Measurement Information System (PROMIS) Roadmap initiative sponsored by the National Institutes of Health with the goal of bringing a more systematic approach to the measurement of health-related variables. I had the privilege of serving as an investigator and chair of the Social Health Domain during the first cycle of PROMIS. The tools developed by this initiative are an important resource to health researchers, and I point readers toward publications that describe it in greater detail. I also highlight papers arising from PROMIS that describe measurement research methods, such as qualitative item review procedures, that readers may find useful.

In summary, this edition updates and expands the topics covered, trims a small amount of material that has proven less helpful than I initially hoped, and discusses new topics that have emerged since the previous edition. I believe that these changes enhance the clarity and utility of the book and hope that you will agree.

About the Author

Robert F. DeVellis is Research Professor in the Department of Health Behavior and Health Education (School of Public Health) at the University of North Carolina at Chapel Hill. Dr. DeVellis has more than 30 years of experience in the measurement of psychological and social variables. He has been an active member of the Patient-Reported Outcomes Measurement Information System (PROMIS) consortium, a multisite National Institutes of Health (NIH) Roadmap initiative directed at identifying, modifying, testing, and disseminating outcome measures for use by NIH investigators. His role in PROMIS was as network-wide domain chair for Social Outcomes. He has served on the Board of Directors for the American Psychological Association's Division of Health Psychology (38), on the Arthritis Foundation's Clinical/Outcomes/Therapeutics Research Study Section, and on the Advisory Board of the Veterans Affairs Measurement Excellence Initiative. He is the recipient of the 2005 Distinguished Scholar Award from the Association of Rheumatology Health Professionals and has been selected to serve as associate editor of *Arthritis Care and Research.* In addition, he has served as guest editor, guest associate editor, or reviewer for more than two dozen other journals. His current research interests include examining how spousal and other close relationships can mitigate the adverse impact of illnesses and measuring social and behavioral variables related to health and illness. He has served as principal investigator or co-investigator since the early 1980s on a series of research projects funded by the federal government and private foundations.

1

Overview

Measurement is of vital concern across a broad range of social research contexts. For example, consider the following hypothetical situations:

1. A health psychologist faces a common dilemma: The measurement scale she needs apparently does not exist. Her study requires that she have a measure that can differentiate between what individuals *want* to happen and what they *expect* to happen when they see a physician. Her research shows that previous studies used scales that inadvertently confounded these two ideas. No existing scales appear to make this distinction in precisely the way that she would like. Although she could fabricate a few questions that seem to tap the distinction between what one wants and expects, she worries that "made-up" items might not be reliable or valid indicators of these concepts.

2. An epidemiologist is unsure how to proceed. He is performing secondary analyses on a large data set based on a national health survey. He would like to examine the relationship between certain aspects of perceived psychological stress and health status. Although no set of items intended as a stress measure was included in the original survey, several items originally intended to measure other variables appear to tap content related to stress. It might be possible to pool these items into a reliable and valid measure of psychological stress. However, if the pooled items constitute a poor measure of stress, the investigator might reach erroneous conclusions.

3. A marketing team is frustrated in its attempts to plan a campaign for a new line of high-priced infant toys. Focus groups have suggested that parents' purchasing decisions are strongly influenced by the apparent educational relevance of toys of this sort. The team suspects that parents who have high educational and career aspirations for their infants will be more attracted to this new line of toys. Therefore, the team would like to assess these aspirations among a large and geographically dispersed sample of parents. Additional focus groups are judged to be too cumbersome for reaching a sufficiently large sample of consumers.

In each of these situations, people interested in some substantive area have come head to head with a measurement problem. None of these researchers is interested primarily in measurement per se. However, each must find a way to quantify a particular phenomenon before tackling the main research objective. In each case, "off-the-shelf" measurement tools are either inappropriate or unavailable. All the researchers recognize that adopting haphazard measurement approaches runs the risk of yielding inaccurate data. Developing their own measurement instruments seems to be the only remaining option.

Many social science researchers have encountered similar problems. One all-too-common response to these types of problems is reliance on existing instruments of questionable suitability. Another is to assume that newly developed questionnaire items that "look right" will do an adequate measurement job. Uneasiness or unfamiliarity with methods for developing reliable and valid instruments and the inaccessibility of practical information on this topic are common excuses for weak measurement strategies. Attempts at acquiring scale development skills may lead a researcher either to arcane sources intended primarily for measurement specialists or to information too general to be useful. This volume is intended as an alternative to those choices.

GENERAL PERSPECTIVES ON MEASUREMENT

Measurement is a fundamental activity of science. We acquire knowledge about people, objects, events, and processes by observing them. Making sense of these observations frequently requires that we quantify them (i.e., that we measure the things in which we have a scientific interest). The process of measurement and the broader scientific questions it serves interact with each other; the boundaries between them are often imperceptible. This happens, for example, when a new entity is detected or refined in the course of measurement or when the reasoning involved in determining how to quantify a phenomenon of interest sheds new light on the phenomenon itself. For example, Smith, Earp, and DeVellis (1995) investigated women's perceptions of battering. An a priori conceptual model based on theoretical analysis suggested six distinct components to these perceptions. Empirical work aimed at developing a scale to measure these perceptions indicated that, among both battered and nonbattered women, a much simpler conceptualization prevailed: A single concept thoroughly explained how study participants responded to 37 of 40 items administered. This finding suggests that what researchers saw as a complex constellation of variables was actually perceived by women living in the community as a single, broader phenomenon. Thus, in the course of devising a means of measuring women's perceptions about battering, we discovered something new about the structure of those perceptions.

Duncan (1984) argues that the roots of measurement lie in social processes and that these processes and their measurement actually precede science: "All measurement . . . is social measurement. Physical measures are made for social purposes" (p. 35). In reference to the earliest formal social measurement processes, such as voting, census taking, and systems of job advancement, Duncan notes that "their origins seem to represent attempts to meet everyday human needs, not merely experiments undertaken to satisfy scientific curiosity." He goes on to say that similar processes

> can be drawn in the history of physics: the measurement of length or distance, area, volume, weight and time was achieved by ancient peoples in the course of solving practical, social problems; and physical science was built on the foundations of those achievements. (p. 106)

Whatever the initial motives, each area of science develops its own set of measurement procedures. Physics, for example, has developed specialized methods and equipment for detecting subatomic particles. Within the behavioral/social sciences, *psychometrics* has evolved as the subspecialty concerned with measuring psychological and social phenomena. Typically, the measurement procedure used is the questionnaire and the variables of interest are part of a broader theoretical framework.

HISTORICAL ORIGINS OF MEASUREMENT IN SOCIAL SCIENCE

Early Examples

Common sense and the historical record support Duncan's claim that social necessity led to the development of measurement before science emerged. No doubt, some form of measurement has been a part of our species' repertoire since prehistoric times. The earliest humans must have evaluated objects, possessions, and opponents on the basis of characteristics such as size. Duncan (1984) cites biblical references to concerns with measurement (e.g., "A false balance is an abomination to the Lord, but a just weight is a delight," Proverbs 11:1) and notes that the writings of Aristotle refer to officials charged with checking weights and measures. Anastasi (1968) notes that the Socratic method employed in ancient Greece involved probing for understanding in a manner that might be regarded as knowledge testing. In his 1964 essay, P. H. DuBois (reprinted in Barnette, 1976) describes the use of civil service testing as early as 2200 B.C. in China. Wright (1999) cites other examples of the importance ascribed in antiquity to accurate measurement, including the "weight of seven" on which 7th-century Muslim taxation was based. He also notes that some have linked the French Revolution, in part, to peasants being fed up with unfair measurement practices.

The notion that measurement can entail error and that certain steps might be taken to reduce that error is a more recent insight. Buchwald (2006), in his review of measurement discrepancies and their impact on knowledge, notes that, while still in his twenties during the late 1660s and early 1670s, Isaac Newton was apparently the first to use an average of multiple observations. His intent was to produce a more accurate measurement when his observations of astronomical phenomena yielded discrepant values. Interestingly, he did not document the use of averages in his initial reports but concealed his reliance on them for decades. This concealment may have stemmed less from a lack of integrity than from a limited understanding of error and its role in measurement. Commenting on another astronomer's similar disdain for discrepant observations, Alder (2002) argues that even in the late 1700s, concealment of discrepancies in observation "were not only common, they were considered a savant's prerogative. It was error that was seen as a moral failing" (p. 301). Buchwald (2006) makes a similar observation:

> [17th- and early 18th-century scientists'] way of working regarded differences not as the inevitable byproducts of the measuring process itself, but as evidence of failed or inadequate skill. Error in measurement was potentially little different from faulty behavior of any kind: it could have moral consequences, and it had to be managed in appropriate ways. (p. 566)

Astronomers were not the only scientists making systematic observations of natural phenomena in the late 1600s and early 1700s. In the 1660s, John Graunt was compiling birth and death rates from christening and burial records in Hampshire, England. Graunt used an averaging procedure (though not the one in common use today) to summarize his findings. According to Buchwald (2006), Graunt's motivation for this averaging was to capture an ephemeral "true" value. The notion was that the ratio of births to deaths obeyed some law of nature but that unpredictable events that might occur in any given year would mask that fundamental truth. This view of observation as an imperfect window into nature's truths suggests a growing sophistication in how measurement was viewed: In addition to the observer's limitations, other factors could also corrupt empirically gathered information, and some adjustments of those values might more accurately reveal the true nature of the phenomenon of interest.

Despite these early insights, it was a century after Newton's first use of the average before scientists more widely recognized that all measurements were prone to error and that an average would minimize such error (Buchwald, 2006). According to physicist and author Leonard Mlodinow (2008), in the late 18th and early 19th centuries, developments in astronomy and physics forced scientists to approach random error more systematically, which led to the emergence of mathematical statistics. By 1777, Daniel Bernoulli (nephew of the more famous Jakob Bernoulli) compared the distributions of values obtained from astronomical observations to the path of an archer's arrows, clumping around a central point with progressively fewer at increasingly greater distances from that center. Although the theoretical treatment that accompanied that observation was wrong in certain respects, it marks the beginning of a formal analysis of error in measurement (Mlodinow, 2008). Buchwald (2006) argues that a fundamental shortcoming of 18th-century interpretations of measurement error was a failure to distinguish between random and systematic error. Not until the dawning of the next century would a more incisive understanding of randomness emerge. With this growing understanding of randomness came advances in measurement; and, as measurement advanced, so did science.

Emergence of Statistical Methods and the Role of Mental Testing

Nunnally's (1978) perspective supports the view that a more sophisticated understanding of randomness, probability and statistics, was necessary for measurement to flourish. He argues that, although systematic observations may have been going on, the absence of more formal statistical methods hindered the development of a science of measuring human abilities until the latter half of the 19th century. The eventual development of suitable statistical methods in the 19th century was set in motion by Darwin's work on evolution and his observation and measurement of systematic variation across species. Darwin's cousin, Sir Francis Galton, extended the systematic observation of differences to humans. A chief concern of Galton was the inheritance of anatomical and intellectual traits. Karl Pearson, regarded by many as the "founder of statistics" (e.g., Allen & Yen, 1979, p. 3), was a junior colleague of Galton's. Pearson developed the mathematical tools—including the Product-Moment Correlation Coefficient bearing his name—needed to systematically examine relationships among variables. Scientists could then quantify the extent to which measurable characteristics were interrelated. Charles Spearman continued in the tradition of his predecessors and set the stage for the subsequent development and popularization of factor analysis in the early 20th century. It is noteworthy that many of the early contributors to formal measurement (including Alfred Binet, who developed tests of mental ability in France in the early 1900s) shared an interest in intellectual abilities. Hence, much of the early work in psychometrics was applied to "mental testing."

The Role of Psychophysics

Another historical root of modern psychometrics arose from psychophysics. As we have seen, measurement problems were common in astronomy and other physical sciences and were a source of concern for Sir Isaac Newton (Buchwald, 2006). Psychophysics exists at the juncture of psychology and physics and concerns the linkages between the physical properties of stimuli and how they are perceived by humans. Attempts to apply the measurement procedures of physics to the study of sensations led to a protracted debate regarding the nature of measurement. Narens and Luce (1986) have summarized the issues. They note that in the late 19th century, Helmholtz observed that physical attributes, such as length and mass, possessed the same intrinsic mathematical structure as did positive real numbers. For example, units of length or mass could be ordered and added as could ordinary numbers. In the early 1900s, the debate continued. The Commission of the British Association for the Advancement of Science regarded fundamental measurement of psychological variables to be impossible because of the problems inherent in ordering or adding sensory perceptions. S. S. Stevens argued that strict additivity, as would apply to length or mass, was not necessary and pointed out that individuals could make fairly consistent ratio judgments of sound intensity. For example, they could judge one sound to be twice or half as loud as another. He argued that this ratio property enabled the data from such measurements to be subjected to mathematical manipulation. Stevens is credited with classifying measurements into nominal, ordinal, interval, and ratio scales. Loudness judgments, he argued, conformed to a ratio scale (Duncan, 1984). At about the time that Stevens was presenting his arguments on the legitimacy of scaling psychophysical measures, L. L. Thurstone was developing the mathematical foundations of factor analysis (Nunnally, 1978). Thurstone's interests spanned both psychophysics and mental abilities. According to Duncan (1984), Stevens credited Thurstone with applying psychophysical methods to the scaling of social stimuli. Thus, his work represents a convergence of what had been separate historical roots.

LATER DEVELOPMENTS IN MEASUREMENT

Evolution of Basic Concepts

As influential as Stevens has been, his conceptualization of measurement is by no means the final word. He defined measurement as the "assignment of numerals to objects or events according to rules" (Duncan, 1984). Duncan challenged this definition as

> incomplete in the same way that "playing the piano is striking the keys of the instrument according to some pattern" is incomplete. Measurement is not only the assignment of numerals, etc. It is also the assignment of numerals in such a way as to correspond to *different degrees of a quality* . . . or property of some object or event. (p. 126)

Narens and Luce (1986) also identified limitations in Stevens's original conceptualization of measurement and illustrated a number of subsequent refinements. However, their work underscores a basic point made by Stevens: Measurement models other than the type endorsed by the Commission (of the British Association for the Advancement of Science) exist, and these lead to measurement methods applicable to the nonphysical as well as physical sciences. In essence, this work on the fundamental properties of measures has established the scientific legitimacy of the types of measurement procedures used in the social sciences.

Evolution of Mental Testing

Although, traditionally, mental testing (or ability testing, as it is now more commonly known) has been an active area of psychometrics, it is not a primary focus of this volume. Nonetheless, it bears mention as a source of significant contributions to measurement theory and methods. A landmark publication, *Statistical Theories of Mental Test Scores,* by Frederic M. Lord and Melvin R. Novick, first appeared in 1968 and has recently been reissued (Lord & Novick, 2008). This volume grew out of the rich intellectual activities of the Psychometric Research Group of the Educational Testing Service, where Lord and Novick were based. This impressive text summarized much of what was known in the area of ability testing at the time and was among the first cogent descriptions of what has become known as *item response theory*. The latter approach was especially well suited to an area as broad as mental testing. Many of the advances in that branch of psychometrics are less common, and perhaps less easily applied, when the goal is to measure characteristics other than mental abilities. Over time, the applicability of these methods to measurement contexts other than ability assessment has become more apparent, and we will discuss them in a later chapter. Primarily, however, I will emphasize the "classical" methods that largely have dominated the measurement of social and psychological phenomena other than abilities. These methods are generally more tractable for nonspecialists and can yield excellent results.

Broadening the Domain of Psychometrics

Duncan (1984) notes that the impact of psychometrics in the social sciences has transcended its origins in the measurement of sensations and intellectual abilities. Psychometrics clearly has emerged as a methodological paradigm in its own right. Duncan supports this argument with three examples of the impact of psychometrics: (1) the widespread use of psychometric definitions of reliability and validity, (2) the popularity of factor analysis in social science research, and (3) the adoption of psychometric methods for developing scales measuring an array of variables far broader than those with which psychometrics was initially concerned (p. 203). The applicability of psychometric concepts and methods to the measurement of diverse psychological and social phenomena will occupy our attention for the remainder of this volume.

THE ROLE OF MEASUREMENT IN THE SOCIAL SCIENCES

The Relationship of Theory to Measurement

The phenomena we try to measure in social science research often derive from theory. Consequently, theory plays a key role in how we conceptualize our measurement problems. In fact, Lord and Novick (2008) ascribe theoretical issues an important role in the development of measurement theory. Theoreticians were concerned that estimates of relationships between constructs of interest were generally obtained by correlating *indicators* of those constructs. Because those indicators contained error, the resultant correlations were an underestimate of the actual relationship between the constructs. This motivated the development of methods of adjusting correlations for error-induced attenuation and stimulated the development of measurement theory as a distinct area of concentration (p. 69).

Of course, many areas of science measure things derived from theory. Until a subatomic particle is confirmed through measurement, it too is merely a theoretical construct. However, theory in psychology and other social sciences is different from theory in the physical sciences. Social scientists tend to rely on numerous theoretical models that concern rather narrowly circumscribed phenomena, whereas theories in the physical sciences are fewer in number and more comprehensive in scope. Festinger's (1954) social comparison theory, for example, focuses on a rather narrow range of human experience: the way people evaluate their own abilities or opinions by comparing themselves with others. In contrast, physicists continue to work toward a grand unified field theory that will embrace all the fundamental forces of nature within a single conceptual framework. Also, the social sciences are less mature than the physical sciences, and their theories are evolving more rapidly. Measuring elusive, intangible phenomena derived from multiple, evolving theories poses a clear challenge to social science researchers. Therefore, it is especially important to be mindful of measurement procedures and to fully recognize their strengths and shortcomings.

The more researchers know about the phenomena in which they are interested, the abstract relationships that exist among hypothetical constructs, and the quantitative tools available to them, the better equipped they are to develop reliable, valid, and usable scales. Detailed knowledge of the specific phenomenon of interest is probably the most important of these considerations. For example, social comparison theory has many aspects that may imply different measurement strategies. One research question might require operationalizing social comparisons as relative preference for information about higher- or lower-status others, while another might dictate ratings of self relative to the "typical person" on various dimensions. Different measures capturing distinct aspects of the same general phenomenon (e.g., social comparison) thus may not yield convergent results (DeVellis et al., 1990). In essence, the measures are assessing different variables despite the use of a common variable name in their descriptions. Consequently, developing a measure that is optimally suited to the research question requires understanding the subtleties of the theory.

Different variables call for different assessment strategies. Number of tokens taken from a container, for example, can be observed directly. Many—arguably, most—of the variables of interest to social and behavioral scientists are not directly observable; beliefs, motivational states, expectancies, needs, emotions, and social role perceptions are but a few examples. Certain variables cannot be directly observed but can be determined by research procedures other than questionnaires. For example, although cognitive researchers cannot directly observe how individuals organize information about gender into their self schemas, they may be able to use recall procedures to make inferences about how individuals structure their thoughts about self and gender. There are many instances, however, in which it is impossible or impractical to assess social science variables with any method other than a paper-and-pencil measurement scale. This is often, but not always, the case when we are interested in measuring theoretical constructs. Thus, an investigator interested in measuring androgyny may find it far easier to do so by means of a carefully developed questionnaire than by some alternative procedure.

Theoretical and Atheoretical Measures

At this point, we should acknowledge that although this book focuses on measures of theoretical constructs, not all paper-and-pencil assessments need be theoretical. Sex and age, for example, can be ascertained from self-report by means of a questionnaire. Depending on the research question, these two variables can be components of a theoretical model or simply part of a description of a study's participants. Some contexts in which people are asked to respond to a list of questions using a paper-and-pencil format, such as an assessment of hospital patient meal preferences, have no theoretical foundation. In other cases, a study may begin atheoretically but result in the formulation of theory. For example, a market researcher might ask parents to list the types of toys they have bought for their children. Subsequently, the researcher might explore these listings for patterns of relationships. Based on the observed patterns of toy purchases, the researcher may develop a model of purchasing behavior. Public opinion questionnaires are another example of relatively atheoretical measurement. Asking people which brand of soap they use or for whom they intend to vote seldom involves any attempt to tap an underlying theoretical construct. Rather, the interest is in the subject's response per se, not in some characteristic of the person it is presumed to reflect.

Distinguishing between theoretical and atheoretical measurement situations can be difficult at times. For example, seeking a voter's preference in presidential candidates as a means of predicting the outcome of an election amounts to asking a respondent to report his or her behavioral intention. An investigator may ask people how they plan to vote not out of an interest in voter decision-making processes but merely to anticipate the eventual election results. If, on the other hand, the same question is asked in the context of examining how attitudes toward specific issues affect candidate preference, a well-elaborated theory may underlie the research. The information about voting is not intended in this case to reveal how the respondent will vote but to shed light on individual characteristics. In these two instances the relevance or irrelevance of the measure to theory is a matter of the investigator's intent, not the procedures used. Readers interested in learning more about constructing survey questionnaires that are not primarily concerned with measuring hypothetical constructs are referred to Converse and Presser (1986); Czaja and Blair (1996); Dillman (2007); Fink (1995); Fowler (2009); and Weisberg, Krosnick, and Bowen (1996).

Measurement Scales

Measurement instruments that are collections of items combined into a composite score and intended to reveal levels of theoretical variables not readily observable by direct means are often referred to as *scales*. We develop scales when we want to measure phenomena that we believe to exist because of our theoretical understanding of the world but that we cannot assess directly. For example, we may invoke depression or anxiety as explanations for behaviors we observe. Most theoreticians would agree that depression or anxiety is not equivalent to the behavior we see but underlies it. Our theories suggest that these phenomena exist and that they influence behavior but that they are intangible. Sometimes, it may be appropriate to infer their existence from their behavioral consequences. However, at other times, we may not have access to behavioral information (as when we are restricted to mail survey methodologies), may not be sure how to interpret available samples of behavior (as when a person remains passive in the face of an event that most others would react to strongly), or may be unwilling to assume that behavior is isomorphic with the underlying construct of interest (as when we suspect that crying is the result of joy rather than sadness). In instances when we cannot rely on behavior as an indication of a phenomenon, it may be more useful to assess the construct by means of a carefully constructed and validated scale.

Even among theoretically derived variables, there is an implicit continuum ranging from relatively concrete and accessible to relatively abstract and inaccessible phenomena. Not all will require multi-item scales. Age and gender certainly have relevance to many theories but rarely require a multi-item scale for accurate assessment. People know their age and gender. These variables, for the most part, are linked to concrete, relatively unambiguous characteristics (e.g., morphology) or events (e.g., date of birth). Unless some special circumstance such as a neurological impairment is present, respondents can retrieve information about their age and gender from memory quite easily. They can respond with a high degree of accuracy to a single question assessing a variable such as these. Ethnicity arguably is more complex and abstract than age or gender. It typically involves a combination of physical, cultural, and historical factors. As a result, it is less tangible—more of a social construction—than age or gender. Although the mechanisms involved in defining one's ethnicity may be complex and unfold over an extended period of time, most individuals have arrived at a personal definition and can report their ethnicity with little reflection or introspection. Thus, a single variable may suffice for assessing ethnicity under most circumstances. (This may change, however, as our society becomes progressively more multiethnic and as individuals define their personal ethnicity in terms of multiple ethnic groups reflecting their ancestry.) Many other theoretical variables, however, require a respondent to reconstruct, interpret, judge, compare, or evaluate less accessible information. For example, measuring how married people believe their lives would be different if they had chosen a different spouse probably would require substantial mental effort, and one item may not capture the complexity of the phenomenon of interest. Under conditions such as these, a scale may be a more appropriate assessment tool. Multiple items may capture the essence of such a variable with a degree of precision that a single item could not attain. It is precisely this type of variable—one that is not directly observable and that involves thought on the part of the respondent—that is most appropriately assessed by means of a scale.

A scale should be contrasted to other types of multi-item measures that yield a composite score. The distinctions among these different types of item composites are of both theoretical and practical importance, as later chapters will reveal. As the terms are used in this volume, a scale consists of what Bollen (1989, pp. 64–65; see also Loehlin, 1998, pp. 200–202) refers to as "effect indicators"—that is, items whose values are caused by an underlying construct (or *latent variable*, as we shall refer to it in the next chapter). A measure of depression often conforms to the characteristics of a scale, with the responses to individual items sharing a common cause—namely, the affective state of the respondent. Thus, how someone responds to items such as "I feel sad" and "My life is joyless" probably is largely determined by that person's feelings at the time. I will use the term *index,* on the other hand, to describe sets of items that are cause indicators—that is, items that determine the level of a construct. A measure of presidential candidate appeal, for example, might fit the characteristics of an index. The items might assess a candidate's geographical residence, family size, physical attractiveness, ability to inspire campaign workers, and potential financial resources. Although these characteristics probably do not share any common cause, they might all share an effect—increasing the likelihood of a successful presidential campaign. The items are not the result of any one thing, but they determine the same outcome. A more general term for a collection of items that one might aggregate into a composite score is *emergent variable* (e.g., Cohen, Cohen, Teresi,

Marchi, & Velez, 1990), which includes collections of entities that share certain characteristics and can be grouped under a common category heading. Grouping them together, however, does not necessarily imply any causal linkage. Sentences beginning with a word having fewer than five letters, for example, can easily be categorized together although they share neither a common cause nor a common effect. An emergent variable "pops up" merely because someone or something (such as a data analytic program) perceives some type of similarity among the items in question.

All Scales Are Not Created Equal

Regrettably, not all item composites are developed carefully. For many, *assembly* may be a more appropriate term than *development*. Researchers often throw together or dredge up items and assume they constitute a suitable scale. These researchers may give no thought to whether the items share a common cause (thus constituting a scale), share a common consequence (thus constituting an index), or merely are examples of a shared superordinate category that does not imply either a common causal antecedent or consequence (thus constituting an emergent variable).

A researcher not only may fail to exploit theory in developing a scale but also may reach erroneous conclusions about a theory by misinterpreting what a scale measures. An unfortunate but distressingly common occurrence is the conclusion that some *construct* is unimportant or that some *theory* is inconsistent based on the performance of a *measure* that may not reflect the variable assumed by the investigator. Why might this happen? Rarely in research do we directly examine relationships among variables. As noted earlier, many interesting variables are not directly observable, a fact we can easily forget. More often, we assess relationships among proxies (such as scales) that are intended to represent the variables of interest. The observable proxy and the unobservable variable may become confused. For example, variables such as blood pressure and body temperature, at first consideration, appear to be directly observable, but what we actually observe are proxies, such as a column of mercury. Our conclusions about the variables assume that the observable proxies are closely linked to the underlying variables they are intended to represent. Such is the case for a thermometer; we describe the level of the mercury in a thermometer as "the temperature" even though, strictly speaking, it is merely a visible manifestation of temperature (i.e., thermal energy). In this case, where the two closely correspond, the consequences of referring to the measurement (scale value that the mercury attains) as the variable (amount of thermal energy) are nearly always inconsequential. When the relationship between the variable and its indicator is weaker than in the thermometer example, confusing the measure with the phenomenon it is intended to reveal can lead to erroneous conclusions. Consider a hypothetical situation in which an investigator wishes to perform a secondary analysis on an existing data set. Let us assume that our investigator is interested in the role of social support on subsequent professional attainment. The investigator observes that the available data set contains a wealth of information on subjects' professional statuses over an extended period of time and that subjects were asked whether they were married. In fact, there may be several items, collected at various times, that pertain to marriage. Let us further assume that, in the absence of any data providing a more detailed assessment of social support, the investigator decides to sum these marriage items into a "scale" and to use this as a measure of support. Most social scientists would agree that equating social support with marital status is not justified. The latter both omits important aspects of social support (e.g., the perceived quality of support received) and includes potentially irrelevant factors (e.g., status as a child too young to have married versus an adult of an age suitable for marriage at the time of measurement). If this hypothetical investigator concluded, on the basis of this assessment method, that social support played no role in professional attainment, that conclusion might be completely wrong. In fact, the comparison was between marital status and professional attainment (or, more precisely, indicators of these variables). Only if marriage actually indicated level of support would the conclusion about support and professional attainment be valid.

Costs of Poor Measurement

Even if a poor measure is the only one available, the costs of using it may be greater than any benefits attained. Situations are rare in the social sciences in which an immediate decision must be made in order to avoid dire consequences and one has no other choice but to make do with the best instruments available. Even in these rare instances, however, the inherent problems of using poor measures to assess constructs do not vanish. Using a measure that does not assess what one presumes can lead to wrong decisions. Does this mean that we should use only measurement tools that have undergone rigorous development and extensive validation testing? Although imperfect measurement may be better than no measurement at all in some situations, we should *recognize* when our measurement procedures are flawed and temper our conclusions accordingly.

Often, an investigator will consider measurement as secondary to more important scientific issues that motivate a study and, thus, the researcher will attempt to economize by skimping on measurement. However, adequate measures are a necessary condition for valid research. Investigators should strive for an isomorphism between the theoretical constructs in which they have an interest and the methods of measurement they use to operationalize them. Poor measurement imposes an absolute limit on the validity of the conclusions one can reach. For an investigator who prefers to pay as little attention to measurement and as much to substantive issues as possible, an appropriate strategy might be to get the measurement part of the investigation correct from the very beginning so that it can be taken more or less for granted thereafter.

A researcher also can falsely economize by using scales that are too brief in the hope of reducing the burden on respondents. Choosing a questionnaire that is too brief to be reliable is a bad idea no matter how much respondents prefer its brevity. A reliable questionnaire that is completed by half of the respondents yields more information than an unreliable questionnaire completed by all respondents. If you cannot determine what the data mean, the amount of information collected is irrelevant. Consequently, completing "convenient" questionnaires that cannot yield meaningful information is a poorer use of respondents' time and effort than completing a somewhat longer version that produces valid data. Thus, using inadequately brief assessment methods may have ethical as well as scientific implications.

SUMMARY AND PREVIEW

This chapter stresses that measurement is a fundamental activity in all branches of science, including the behavioral and social sciences. Psychometrics, the specialty area of the social sciences that is concerned with measuring social and psychological phenomena, has historical antecedents extending back to ancient times. In the social sciences, theory plays a vital role in the development of measurement scales, which are collections of items that reveal the level of an underlying theoretical variable. However, not all collections of items constitute scales in this sense. Developing scales may be more demanding than selecting items casually; however, the costs of using casually constructed measures usually greatly outweigh the benefits.

The following chapters cover the rationale and methods of scale development in greater detail. Chapter 2 explores the latent variable, the underlying construct that a scale attempts to quantify, and presents the theoretical bases for the methods described in later chapters. Chapter 3 provides a conceptual foundation for understanding reliability and the logic underlying the reliability coefficient. Chapter 4 reviews validity, while Chapter 5 is a practical guide to the steps involved in scale development. Chapter 6 introduces factor analytic concepts and describes their use in scale development. Chapter 7 is a conceptual overview of an alternative approach to scale development—item response theory. Finally, Chapter 8 briefly discusses how scales fit into the broader research process.

2

Understanding the Latent Variable

This chapter presents a conceptual schema for understanding the relationship between measures and the constructs they represent, though it is not the only framework available. Item response theory is an alternative measurement perspective that we will examine in Chapter 7. Because of its relative conceptual and computational accessibility and wide usage, I emphasize the classical measurement model, which assumes that individual items are comparable indicators of the underlying construct.

CONSTRUCTS VERSUS MEASURES

Typically, researchers are interested in constructs rather than items or scales per se. For example, a market researcher measuring parents' aspirations for their children would be more interested in intangible parental sentiments and hopes about what their children will accomplish than in where those parents place marks on a questionnaire. However, recording responses to a questionnaire may, in many cases, be the best method of assessing those sentiments and hopes. Scale items are usually a means to the end of construct assessment. In other words, they are necessary because many constructs cannot be assessed directly. In a sense, measures are proxies for variables that we cannot directly observe. By assessing the relationships between measures, we indirectly infer the relationships between constructs. In Figure 2.1, for example, although our primary interest is the relationship between Variables A and B, we estimate that relationship on the basis of the connection between measures corresponding to those variables.

The underlying phenomenon or construct that a scale is intended to reflect is often called the *latent variable.* Exactly what is a latent variable? Its name reveals two chief features. Consider the example of parents' aspirations for children's achievement. First, it is *latent* rather than manifest. Parents' aspirations for their children's achievement are not directly observable. In addition, the construct is *variable* rather than constant—that is, some aspect of it, such as its strength or magnitude, changes. Parents' aspirations for their children's achievement may vary according to time (e.g., during the child's infancy versus adolescence), place (e.g., on an athletic

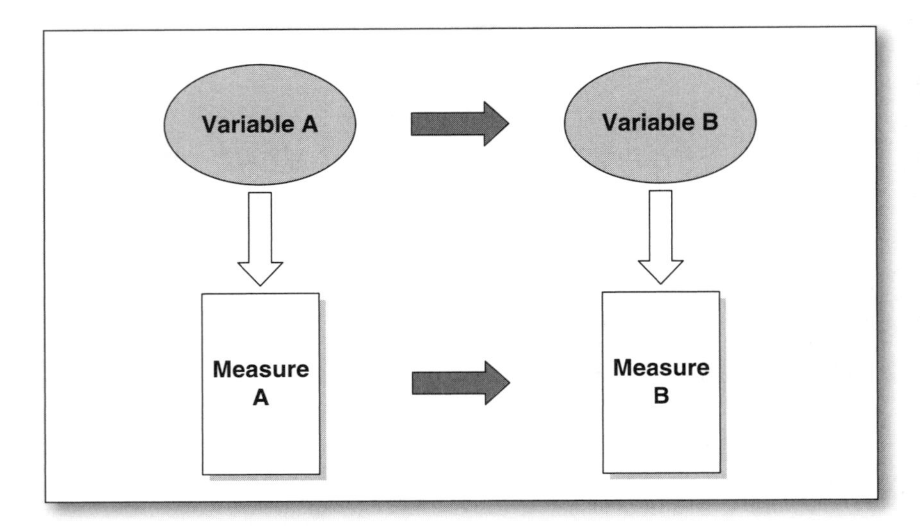

Figure 2.1 Relationships between instruments correspond with relationships between latent variables only when each measure corresponds to its latent variable

field versus a classroom), people (e.g., parents whose own backgrounds or careers differ), or any combination of these and other dimensions. The latent variable is the actual phenomenon that is of interest—in this case, child achievement aspirations.

Another noteworthy aspect of the latent variable is that it is typically a characteristic of the individual who is the source of data. Thus, in our present example, parental aspirations are a characteristic of the parents and not of the children. Accordingly, we assess it by collecting data about the parents' beliefs from the parents themselves. While there may be circumstances in which some form of proxy reporting (e.g., asking parents to report some characteristic of their children) is appropriate, in general, we will ask respondents to self-report information pertaining to themselves. When this is not the case, as in a study involving parents describing the aspirations their children have for themselves, care must be taken in interpreting the resulting information. Arguably, in this hypothetical instance, the latent variable might more accurately be described as *parents' perceptions of their children's aspirations* than as *children's aspirations* per se. Likewise, if we ask a group of shoppers to evaluate characteristics of a particular store, we are assessing *shoppers' perceptions* rather than aspects of the store itself (which might be more easily assessed by direct observation). How important the distinction is between assessing the perceptions of a respondent with regard to some external stimulus (e.g., perceptions of the store), as opposed to characteristics of the external stimulus (e.g., the store itself), will depend on the specific circumstances and goals of the assessment; however, in all cases, it is important to be mindful of the distinction and to make appropriate interpretations of the resultant data.

Although we cannot observe or quantify it directly, the latent variable presumably takes on a specific value under some specified set of conditions. A scale developed to measure a latent variable is intended to estimate its actual magnitude at the time and place of measurement for each thing measured. This unobservable actual magnitude is the *true score.*

LATENT VARIABLE AS THE PRESUMED CAUSE OF ITEM VALUES

The notion of a latent variable implies a certain relationship between it and the items that tap it. The latent variable is regarded as a *cause* of the item score—that is, the strength or quantity of the latent variable (i.e., the value of its true score) is presumed to cause an item (or set of items) to take on a certain value.

An example may reinforce this point: The following are hypothetical items for assessing parents' aspirations for children's achievement:

1. My child's achievements determine my own success.
2. I will do almost anything to ensure my child's success.
3. No sacrifice is too great if it helps my child achieve success.
4. My child's accomplishments are more important to me than just about anything else I can think of.

If parents were given an opportunity to express how strongly they agree with each of these items, their underlying aspirations for childhood achievement should influence their responses. In other words, each item should give an indication of how strong the latent variable (aspirations for children's achievement) is. The score obtained on the item is caused by the strength or quantity of the latent variable for that person at that particular time.

A causal relationship between a latent variable and a measure implies certain empirical relationships. For example, if an item value is caused by a latent variable, then there should be a correlation between that value and the true score of the latent variable. As a consequence of each of the indicators correlating with the latent variable, they should also correlate with each other. Because we cannot directly assess the true score, we cannot compute a correlation between it and the item. However, when we examine a set of items that are presumably caused by the same latent variable, we can examine their relationships to one another. So, if we had several items like the ones above measuring parental aspirations for child achievement, we could look directly at how they correlated with one another, invoke the latent variable as the basis for the correlations among items, and use that information to infer how highly each item was correlated with the latent variable. Shortly, I will explain how all this can be learned from correlations among items. First, however, I will introduce some diagrammatic procedures to help make this explanation more clear.

PATH DIAGRAMS

Coverage of this topic will be limited to a brief review of issues pertinent to scale development. For greater depth, consult Asher (1983) or Loehlin (1998).

Diagrammatic Conventions

Path diagrams are a method for depicting *causal* relationships among variables. Although they can be used in conjunction with path analysis, which is a data analytic method, path diagrams have more general utility as a means of specifying how a set of variables are interrelated. These diagrams adhere to certain conventions. A *straight arrow* drawn from one variable label to another indicates that the two are *causally related* and that the direction of causality is as indicated by the arrow. Thus X ⟶ Y indicates explicitly that X is the cause of Y. Often, associational paths are identified by labels, such as the letter *a* in Figure 2.2.

The *absence* of an arrow also has an explicit meaning—namely, that two variables are *unrelated.* Thus, A ⟶ B ⟶ C D ⟶ E specifies that A causes B, B causes C, C and D are *unrelated,* and D causes E.

Another convention of path diagrams is the method of representing *error,* which is usually depicted as an additional causal variable. This error term is a *residual,* representing all sources of variation not accounted for by other causes explicitly depicted in the diagram.

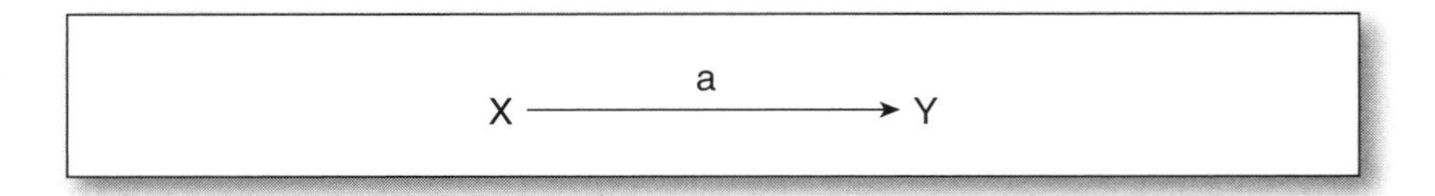

Figure 2.2 The causal pathway from X to Y

Because this error term is a residual, it represents the discrepancy between the actual value of Y and what we would predict Y to be based on knowledge of X and Z (in this case; see Figure 2.3). Sometimes, the error term is assumed and, thus, not included in the diagram.

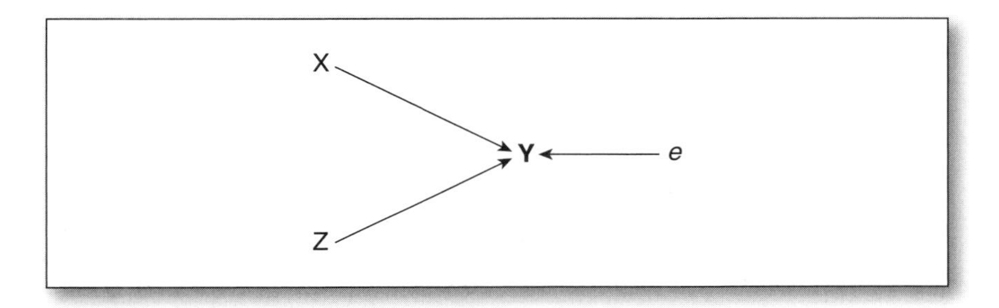

Figure 2.3 Two variables plus error determine Y

Path Diagrams in Scale Development

Path diagrams can help us see how scale items are causally related to a latent variable. They can also help us understand how certain relationships among items imply certain relationships between items and the latent variable. We begin by examining a simple computational rule for path diagrams. Let us look at the simple path diagram in Figure 2.4.

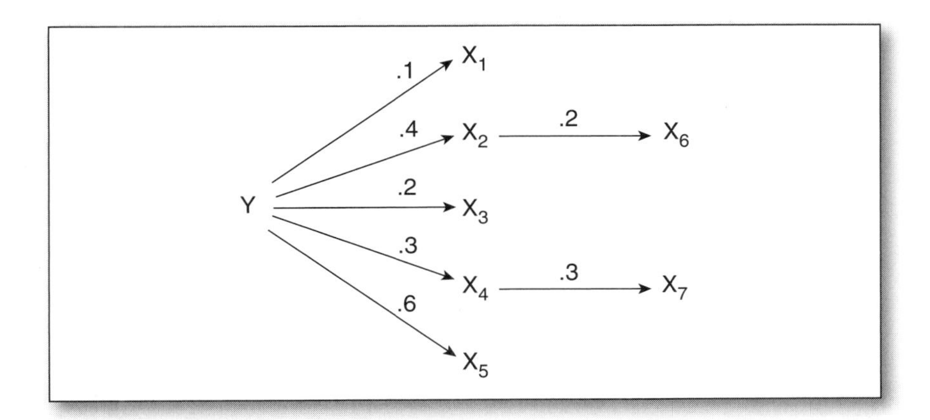

Figure 2.4 A path diagram with path coefficients, which can be used to compute correlations between variables

The numbers along the paths are *standardized path coefficients.* Each one expresses the strength of the causal relationship between the variables joined by the arrow. The fact that the coefficients are standardized means that they all use the same scale to quantify the causal relationships. In this diagram, Y is a cause of X_1 through X_5. A useful relationship exists between the values of path coefficients and the correlations between the Xs (which would represent items in the case of a scale-development–type path diagram). For diagrams like this one having only one common origin (Y in this case), the correlation between any two Xs is equal to the product of the coefficients for the arrows forming a route, through Y, between the X variables in question. For example, the correlation between X_1 and X_5 is calculated by multiplying the two standardized path coefficients that join them via Y. Thus, $r_{1,5} = .6 \times .1 = .06$. Variables X_6 and X_7 also share Y as a common source, but the route connecting them is longer. However, the rule still applies. Beginning at X_7, we can trace back to Y and then forward again to X_6 (or in the other direction, from X_6 to X_7). The result is $.3 \times .3 \times .4 \times .2 = .0072$. Thus, $r_{6,7} = .0072$.

This relationship between path coefficients and correlations provides a basis for estimating paths between a latent variable and the items that it influences. Even though the latent variable is hypothetical and unmeasurable, the items are real and the correlations among them can be directly computed. By using these correlations, the simple rule just discussed, and some assumptions about the relationships among items and the true score, we can come up with estimates for the paths between the items and the latent variable. We can begin with a set of correlations among variables. Then, working backward from the relationship among paths and correlations, we can determine what the values of certain paths must be if the assumptions are correct. Let us consider the example in Figure 2.5.

This diagram is similar to the example considered earlier in Figure 2.4 except that there are no path values, the variables X_6 and X_7 have been dropped, the remaining X variables represent scale items, and each item has a variable (error) other than Y influencing it. These *e* variables are unique in the case of each item and represent the residual variation in each item not explained by Y. This diagram indicates that all the items are influenced by Y. In addition, each is influenced by a unique set of variables other than Y that are collectively treated as error.

This revised diagram represents how five individual items are related to a single latent variable, Y. The numerical subscripts given to the *e*s and Xs indicate that the five items are different and that the five sources of error, one for each item, are also different. The diagram has no arrows going directly from one X to another X or going from an *e* to another *e* or from an *e* to an X other than the one with which it is associated. These aspects of the diagram represent assumptions that will be discussed later.

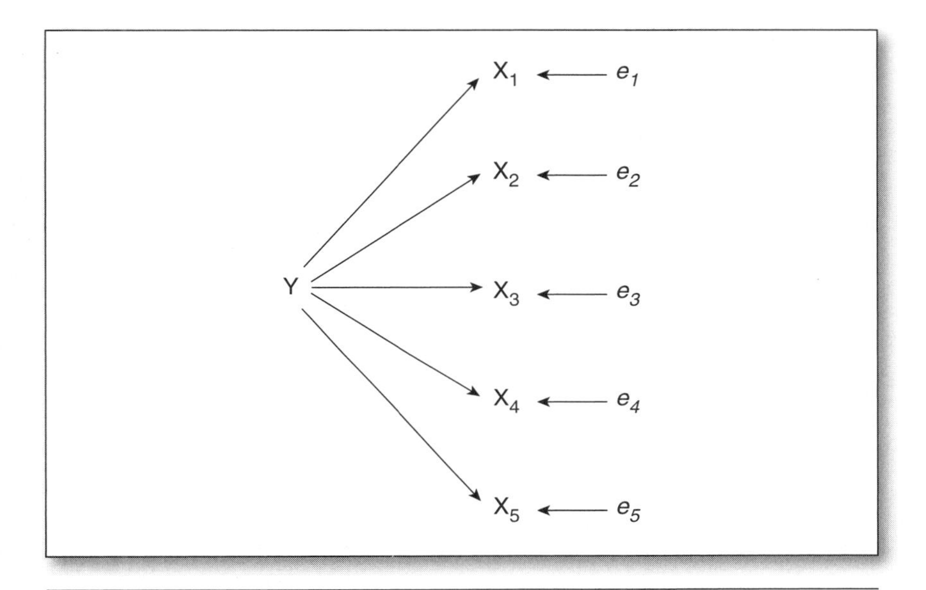

Figure 2.5 A path diagram with error terms

If we had five actual items that a group of people had completed, we would have item scores that we could then correlate with one another. The rule examined earlier allowed the computations of correlations from path coefficients. With the addition of some assumptions, it also lets us compute path coefficients from correlations—that is, correlations computed from actual items can be used to determine how each item relates to the latent variable. If, for example, X_1 and X_4 have a correlation of .49, then we know that the product of the values for the path leading from Y to X_1 and the path leading from Y to X_4 is equal to .49. We know this because our rule established that the correlation of two variables equals the product of the path coefficients along the route that joins them. If we also assume that the *two path values are equal,* then they both must be .70[1].

FURTHER ELABORATION OF THE MEASUREMENT MODEL

Classical Measurement Assumptions

The classical measurement model—which asserts that an observed score, X, results from the summation of a true score, T, plus error, *e*—starts with common assumptions about items and their relationships to the latent variable and sources of error:

1. The amount of error associated with individual items varies randomly. The error associated with individual items has a mean of zero when aggregated across a large number of people. Thus, items' means tend to be unaffected by error when a large number of respondents complete the items.

2. One item's error term is *not* correlated with another item's error term; the only routes linking items always pass through the latent variable, never through any error term.

3. Error terms are *not* correlated with the true score of the latent variable. Note that the paths emanating from the latent variable do not extend outward to the error terms. The arrow between an item and its error term aims the other way.

The first two assumptions above are common statistical assumptions that underlie many analytic procedures. The third amounts to defining "error" as the residual remaining after considering all the relationships between a set of predictors and an outcome or, in this case, a set of items and their latent variable.

PARALLEL TESTS

Classical measurement theory, in its most orthodox form, is based on the assumption of parallel tests. The term *parallel tests* stems from the fact that one can view each individual item as a "test" for the value of the latent variable. For our purposes, referring to parallel items would be more accurate. However, I will defer to convention and use the traditional name.

A virtue of the parallel tests model is that its assumptions make it quite easy to reach useful conclusions about how individual items relate to the latent variable based on our observations of how the items relate to one another. Earlier, I suggested that, with knowledge of the correlations among items and with certain assumptions, one could make inferences about the paths leading from a causal variable to an item. As will be shown in the next chapter, being able to assign a numerical value to the relationships between the latent variable and the items themselves is quite important. Thus, in this section, I will examine in some detail how the assumptions of parallel tests lead to certain conclusions that make this possible.

The rationale underlying the model of parallel tests is that each item of a scale is precisely as good a measure of the latent variable as any other of the scale items. The individual items are thus *strictly parallel,* which is to say that

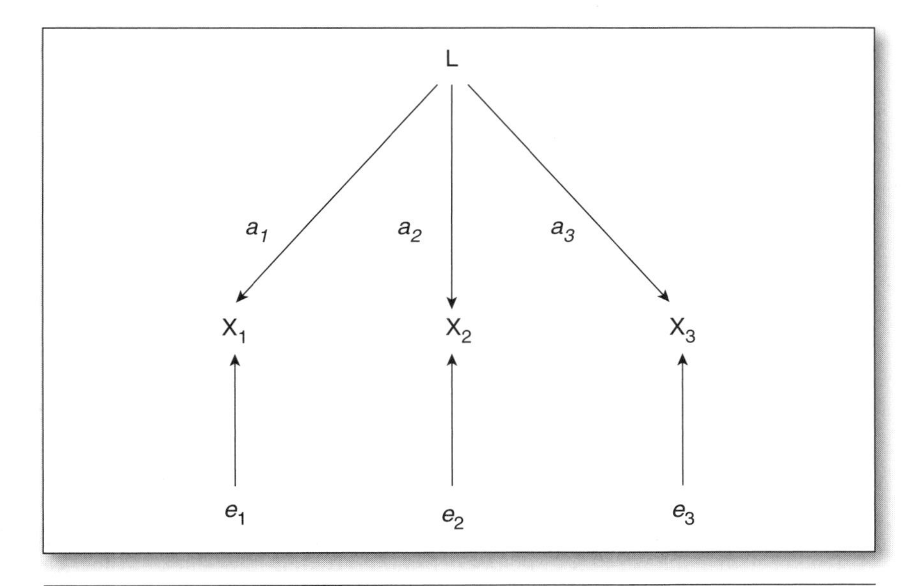

Figure 2.6 A diagram of a parallel tests model, in which all pathways from the latent variable (L) to the items (X_1, X_2, X_3) are equal in value to one another, as are all pathways from the error terms to the items

each item's relationship to the latent variable is presumed identical to every other item's relationship to that variable *and* the amount of error present in each item is also presumed to be identical. Diagrammatically, this model can be represented as shown in Figure 2.6.

This model adds two assumptions to those listed earlier:

1. The amount of influence from the latent variable to each item is assumed to be the same for all items.
2. Each item is assumed to have the same amount of error as any other item, meaning that the influence of factors *other* than the latent variable is equal for all items.

These added assumptions mean that the correlations of each item with the true score are identical. Being able to assert that these correlations are *equal* is important because it leads to a means of determining the *value* for each of these identical correlations. This, in turn, leads to a means of quantifying reliability, which will be discussed in the next chapter.

Asserting that correlations between the true score and each item are equal requires *both* of the preceding assumptions. A squared correlation is the proportion of variance shared between two variables. So, if correlations between the true score and each of two items are equal, the proportions of variance shared between the true score and each item also must be equal. Assume that a true score contributes the same *amount* of variance to each of two items. This amount can be an equal *proportion* of total variance for each item only if the items have identical total variances. In order for the total variances to be equal for the two items, the amount of variance each item receives from sources other than the true score must also be equal. As all variation sources other than the true score are lumped together as error, this means that the two items must have equal error variances. For example, if X_1 got 9 arbitrary units of variation from its true score and 1 from error, the true score proportion would be 90% of total variation. If X_2 also got 9 units of variation from the true score, these 9 units could be 90% of the total only if the total variation were 10. The total could equal 10 only if error contributed 1 unit to X_2 as it did to X_1. The correlation between each item and the true score then would equal the square root of the proportion of each item's variance that is attributable to the true score, or roughly .95 in this case.

Thus, because the parallel tests model assumes that the amount of influence from the latent variable is the same for each item *and* that the amount from other sources (error) is the same for each item, the proportions of item variance attributable to the latent variable and to error are equal for all items. This also means that, under the assumptions of parallel tests, standardized path coefficients from the latent variable to each item are equal for all items. It was assuming that standardized path coefficients were equal that made it possible, in an earlier example, to compute path coefficients from correlations between items. The path diagram rule relating path coefficients to correlations, discussed earlier, should help us understand why these equalities hold when one accepts the preceding assumptions.

The assumptions of this model also imply that correlations among items are identical (e.g., the correlation between X_1 and X_2 is identical to the correlation between X_1 and X_3 or X_2 and X_3). How do we arrive at this conclusion from the assumptions? The correlations are all the same because the only mechanism to account for the correlation between any two items is the route through the latent variable that links those items. For example, X_1 and X_2 are linked only by the route made up of paths a_1 and a_2. The correlation can be computed by tracing the route joining the two items in question and multiplying the path values. For any two items, this entails multiplying two paths that have identical values (i.e., $a_1 = a_2 = a_3$). Correlations computed by multiplying equal values will, of course, be equal.

The assumptions also imply that each of these correlations between items equals the square of any path from the latent variable to an individual item.

How do we reach this conclusion? The product of two different paths (e.g., a_1 and a_2) is identical to the square of either path because both path coefficients are identical. If $a_1 = a_2 = a_3$ and $(a_1 \times a_2) = (a_1 \times a_3) = (a_2 \times a_3)$, then each of these latter products must also equal the value of any of the a paths multiplied by itself. Looking back at Figure 2.6 may make these relationships and their implications clearer.

It also follows from the assumptions of this model that the proportion of error associated with each item is the complement of the proportion of variance that is related to the latent variable. In other words, any effect on a given item that is not explained by the latent variable must be explained by error. Together, these two effects explain 100% of the variation in any given item. This is so simply because the error term (e) is defined as encompassing all sources of variation in the item other than the latent variable.

These assumptions support at least one other conclusion: Because each item is influenced equally by the latent variable and each error term's influence on its corresponding item is also equal, the items all have equal means and equal variances. If the only two sources that can influence the mean are identical for all items, then clearly the means for the items also will be identical. This reasoning also holds for the item variances.

In conclusion, the parallel tests model assumes the following:

1. Random error
2. Errors not correlated with one another
3. Errors not correlated with true score
4. Latent variable affects all items equally
5. Amount of error for each item is equal

These assumptions allow us to reach a variety of interesting conclusions. Furthermore, the model enables us to make inferences about the latent variable based on the items' correlations with one another. However, the model accomplishes this feat by setting forth fairly stringent assumptions.

ALTERNATIVE MODELS

As it happens, all the narrowly restrictive assumptions associated with strictly parallel tests are not necessary in order to make useful inferences about the relationship of true scores to observed scores. A model based on what are technically called *tau-equivalent tests* makes a more liberal assumption—namely, that the amount of error variance associated with a given item need not equal the error variance of the other items (e.g., Allen & Yen, 1979). Tau-equivalent tests still require identical true scores for items, although a slight loosening of that assumption defines *essentially tau-equivalent tests* (or, occasionally, *randomly parallel tests*). Any pair of items adhering to essential tau equivalence may have true scores that differ by some constant. Of course, adding a constant to one item has no effect on any correlation involving that item, because correlations are standardized expressions. Consequently, the correlation between any pair of items or between an item's true score and the item's obtained score is not affected by relaxing the assumptions of strict tau equivalence to those of essential tau equivalence. So what we have said thus far about tau equivalence also applies to essential tau equivalence. In either of these cases, the *standardized* values of the paths from the latent variable to each item may not be equal. However, the *unstandardized* values of the path from the latent variable to each item (i.e., the *amount* as opposed to *proportion* of influence that the latent variable has on each item) are still presumed to be identical for all items. This means that items are parallel with respect to how much they are influenced by the latent variable but are not necessarily influenced to exactly the same extent by extraneous factors that are lumped together as error. Under strictly parallel assumptions, not only do different items tap the true score to the same degree; their error components are also the same. Tau equivalency (tau is the Greek equivalent to *t*, as in true score) is much easier to live with because it does not impose the "equal errors" condition. Because errors may vary, item means and variances may also vary. The more liberal assumptions of this model are attractive because finding equivalent measures of equal variance is rare. This model allows us to reach many of the same conclusions as with strictly parallel tests but with less restrictive assumptions. Readers may wish to compare this model with Nunnally and Bernstein's (1994) discussion of the domain sampling model.

Some scale developers consider even the essentially tau-equivalent model too restrictive. After all, how often can we assume that each item is influenced by the latent variable to the same degree? Tests developed under what is called the *congeneric model* (Jöreskog, 1971) are subject to an even more relaxed set of assumptions (see Carmines & McIver, 1981, for a discussion of congeneric tests). This model assumes (beyond the basic measurement assumptions) merely that all the items share a common latent variable. They need not bear equally strong relationships to the latent variable, and their error variances need not be equal. One must assume only that each item reflects the true score to some degree. Of course, the more strongly each item correlates with the true score, the more reliable the scale will be.

An even less constrained approach is the *general factor model,* which allows multiple latent variables to underlie a given set of items. Carmines and

McIver (1981), Loehlin (1998), and Long (1983) have discussed the merits of this type of very general model, chief among them being its improved correspondence to real-world data. Structural equation modeling approaches often incorporate factor analyses into their measurement models; situations in which multiple latent variables underlie a set of indicators exemplify the general factor model (Loehlin, 1998).

The congeneric model is a special case of the factor model (i.e., a single-factor case). Likewise, an essentially tau-equivalent measure is a special case of a congeneric measure—one for which the relationships of items to their latent variable are assumed to be equal. Finally, a strictly parallel test is a special case of an essentially tau-equivalent one, adding the assumption of equal relationships between each item and its associated sources of error.

Another measurement strategy should be mentioned. This strategy is item response theory (IRT). This approach has been used primarily, but not exclusively, with dichotomous-response (e.g., correct versus incorrect) items in developing ability tests. IRT assumes that each individual item has its own characteristic sensitivity to the latent variable, represented by an item-characteristic curve—a plot of the relationship between the value of the latent variable (e.g., ability) and the probability of a certain response to an item (e.g., answering it correctly). Thus, the curve reveals how much ability an item demands to be answered correctly. We will consider IRT further in Chapter 7.

Except for that consideration of IRT in Chapter 7 and a discussion of factor analysis in Chapter 6, we will focus primarily on parallel and essentially tau-equivalent models for several reasons. First, they exemplify "classical" measurement theory. Second, discussing the mechanisms by which other models operate can quickly become burdensome. Finally, classical models have proven very useful for social scientists with primary interests other than measurement who, nonetheless, take careful measurement seriously. This group is the audience for whom the present text has been written. For these individuals, the scale development procedures that follow from a classical model generally yield satisfactory scales. Indeed, to my knowledge although no tally is readily available, I suspect that (outside ability testing) a substantial majority of the well-known and highly regarded scales used in social science research were developed using such procedures.

EXERCISES

1. How can we infer the relationship between the latent variable and two items related to it based on the correlations between the two items?

2. What is the chief difference in assumptions between the parallel tests and essentially tau-equivalent models?

3. Which measurement model assumes, beyond the basic assumptions common to all measurement approaches, only that the items share a common latent variable?

NOTE

1. Although –.70 is also an allowable square root of .49, deciding between the positive or negative root is typically of less concern than one would think. As long as all the items can be made to correlate positively with one another (if necessary, by reverse scoring certain items, as discussed in Chapter 5), then the signs of the path coefficients from the latent variable to the individual items will be the same and are arbitrary. Note, however, that giving positive signs to these paths implies that the items indicate more of the construct, whereas negative coefficients would imply the opposite.

3

Reliability

Reliability is a fundamental issue in psychological measurement. Its importance is clear once its meaning is fully understood. As the term implies, a reliable instrument is one that performs in consistent, predictable ways. For a scale to be reliable, the scores it yields must represent some true state of the variable being assessed. In practice, this implies that the score produced by the instrument should not change unless there has been an actual change in the variable the instrument is measuring and, thus, that any observed change in scores can be attributed to actual change in that variable. A perfectly reliable scale would be a reflection of the true score and nothing else. This will seldom be achievable; however, we can gauge how closely we approximate that ideal. The more the score we obtain from a scale represents the true score of the variable and the less it reflects other extraneous factors, the more reliable our scale is. Stated more formally, scale reliability is the proportion of variance attributable to the true score of the latent variable. There are several methods for computing reliability, but all share this fundamental definition.

Although alternative methods for computing reliability may appear to be different, the common underlying definition requires that they be computationally equivalent in some basic and important way. This is indeed the case. All these methods involve estimating the variable's true score and determining what proportion of the obtained scale score that true score represents. Our basic measurement model, described in Chapter 2, suggests that an observed score represents the summation of a true score for the variable being assessed plus error arising from extraneous factors. It follows, then, that we can estimate the true score for the variable by subtracting variance arising from error from the total variance of the observed score obtained from a particular measure. We can then compute reliability as a ratio of the estimated true score to the observed score. Thus,

$$\textit{True Score} = \textit{Observed Score} - \textit{Error}$$

$$\textit{Reliability} = \frac{\textit{True Score}}{\textit{Observed Score}}$$

Methods for estimating error are largely what differentiate alternative formulas for computing reliability. Different methods are tailored to specific types of data, although all share a common conceptual foundation: that reliability is the proportion of variance in an observed score that can be attributed to the true score of the variable being assessed.

METHODS BASED ON THE ANALYSIS OF VARIANCE

One means of estimating error is based on the analysis of variance (ANOVA). This data analytic approach partitions the total variance observed into various sources, primarily those that are of substantive interest (i.e., *signal*) and those that arise from some error source (i.e., *noise*), such as imperfections in sampling participants from a population. Although this is not the approach on which we will focus for assessing the reliability of measurement scales, looking at it in condensed form underscores the continuity across definitions of and approaches to reliability.

Thus, by way of a cursory review, consider a very simple set of observations that involve the temperatures of eight identical objects, four of which are in direct sunlight and four of which are in the shade. (I have specified a small number of observed objects in this example for simplicity.) The objects are identical except for their exposure to the sun; however, the thermometer used to measure their temperatures is a bit suspect and, therefore, is a potential source of error in the observed temperatures. We could assess the extent of that error by recording the temperatures of all eight individual objects and arranging the information in several ways. First, we could summarize information about the objects as a single group by computing an overall sum of squared deviations in object temperatures from the overall mean for all the objects. This value would be the total sum of squares, or SS_T. By dividing the SS_T by the degrees of freedom associated with the entire sample (i.e., $N - 1 = 8 - 1 = 7$), we would obtain the total variance for the objects' temperatures. The following steps isolate subcomponents of that overall variance. We could proceed to estimate the extent to which error affected those scores and, thus, was a subcomponent of the total variance. In the ANOVA framework, this is accomplished by assessing how much variation occurs under identical conditions. In this case, all the objects in the sun are exposed to identical conditions, as are all the objects in the shade. Within each of these two subgroups, the objects themselves are presumed to be identical and the presence or absence of sunlight is identical. So, the only basis for differences in *observed* temperatures should be some form of *error*. Thus, we can examine the variation in temperatures of objects within groups to compute an error sum of squares (SS_E). By subtracting this SS_E from SS_T, we can compute a sum of squares for the effect of sunlight. This last sum of squares is essentially the sum of squares for the true score—that is, an indication of the amount of variation in object temperatures after removing the effect of measurement error. We can then compute the true score variance from this sum of squares. Finally, by computing the ratio between that true score variance and the total variance, we arrive at the proportion of total variance that can be attributed to the *true score* (i.e., the effect of the sun). We could interpret that proportion as the reliability of our measurement of the objects' temperatures.

Note that if all the objects in the sun had identical temperatures and all the objects in the shade had identical, presumably lower, temperatures, the error variance would be 0.0. Thus, nothing would get subtracted from the observed SS_T, the true score variance and total variance would be equal, and the ratio representing the reliability of measuring object temperatures would be 1.0.

I have referred to the ratio arising from the ANOVA example described above as a reliability coefficient, which is correct. More generally, however, a ratio comparing the variance arising from some specific source in an ANOVA design with the total variance is known as an intraclass correlation coefficient, or ICC. Depending on the type and complexity of ANOVA design, there can be several types of ICC that will have various interpretations, not all of which are equivalent to measurement reliability. Although readers may not be as familiar with the ICC as they are with other, more common expressions for reliability, we will see that the logic on which more specialized indicators of reliability are based is identical to the logic of the ICC—that is, both the ICC and other methods of capturing reliability are based on a comparison of some estimate of true score variance with total variance.

CONTINUOUS VERSUS DICHOTOMOUS ITEMS

Although items may have a variety of response formats, we assume in this chapter that item responses consist of multiple-value response options. Dichotomous items (i.e., items having only two response options, such as "yes" and "no," or those having multiple response options that can be classified as "right" versus "wrong") are widely used in ability testing and, to a lesser degree, in other measurement contexts. Examples of dichotomous items include the following:

1. Zurich is the capital of Switzerland. True False
2. What is the value of pi?
   (a) 1.41
   (b) 3.14
   (c) 2.78

Special methods for computing reliability that take advantage of the computational simplicity of dichotomous responses have been developed. General measurement texts such as Nunnally and Bernstein's (1994) cover these methods in some detail. The logic of these methods for assessing reliability largely parallels the more general approach that applies to multipoint, continuous scale items. In fact, in some cases, the approach to assessing reliability for multiresponse items is an extension of an earlier approach developed for dichotomous-response items. In the interest of brevity, this chapter will make only passing reference to reliability assessment methods intended for scales made up of dichotomous items. Some characteristics of this type of scale are discussed in Chapter 5.

INTERNAL CONSISTENCY

Internal consistency reliability, as the name implies, is concerned with the homogeneity of the items within a scale. Scales based on classical measurement models are intended to measure a single phenomenon. As we saw in the preceding chapter, measurement theory suggests that the relationships among items are logically connected to the relationships of items to the latent variable. If the items of a scale have a strong relationship to their latent variable, they will have a strong relationship to one another. Although we cannot directly observe the linkage between items and the latent variable, we can certainly determine whether the items are correlated to one another. A scale is *internally consistent* to the extent that its items are highly intercorrelated. What can account for correlations among items? There are two possibilities: Either items causally affect each other (e.g., Item A causes Item B), or the items share a common cause. Under most conditions, the former explanation is unlikely, leaving the latter as the more obvious choice. Thus, high inter-item correlations suggest that the items are all measuring (i.e., are manifestations of) the same thing. If we make the assumptions discussed in the preceding chapter (particularly the assumption that items do not share sources of error), we also can conclude that strong correlations among items imply strong links between items and the latent variable. Thus, a unidimensional scale or a single dimension of a multidimensional scale should consist of a set of items that correlate well with one another. Multidimensional scales measuring several phenomena—for example, the Multidimensional Health Locus of Control scales (Wallston, Wallston, & DeVellis, 1978)—are really families of related scales; each "dimension" is a scale in its own right.

Coefficient Alpha

Internal consistency is typically equated with Cronbach's (1951) coefficient alpha (α). We will examine alpha in some detail for several reasons. First, it is widely used as a measure of reliability. Second, its connection to the definition of reliability may be less evident than is the case for other measures of reliability (such as the alternate forms methods) discussed later. Consequently, alpha may appear more mysterious than other reliability computation methods to those who are not familiar with its internal workings. Finally, an exploration of the logic underlying the computation of alpha provides a sound basis for comparing how other computational methods capture the essence of what we mean by reliability.

The Kuder-Richardson formula 20—or *KR-20,* as it is more commonly known—is a special version of alpha for items that are dichotomous (see, e.g., Nunnally & Bernstein, 1994). KR-20 and coefficient alpha are equivalent when the items composing a scale are dichotomous. However, as noted earlier, we will concentrate on the more general form that applies to items having multiple response options.

You can think about all the variability in a set of item scores as due to one of two things: (a) actual variation across individuals in the phenomenon that the scale measures (i.e., true variation in the latent variable) or (b) error. This is true because classical measurement models define the phenomenon (e.g., patients' desire for control of interactions with a physician) as the source of all shared variation, and they define error as any remaining, or unshared, variation in scale scores (e.g., a single item's unintended double meaning). Another way to think about this is to regard total variation as having two components: signal (i.e., true differences in patients' desire for control) and noise (i.e., score differences caused by everything *but* true differences in desire for control). Computing alpha, as we shall see, partitions the total variance among the set of items into signal and noise components. The proportion of total variation that is signal equals alpha. Thus, another way to think about alpha is that it equals 1 – error variance or, conversely, that error variance = 1 – alpha.

The Covariance Matrix

To understand internal consistency more fully, it helps to examine the *covariance matrix* of a set of scale items. A covariance matrix for a set of scale items reveals important information about the scale as a whole.

A covariance matrix is a more general form of a correlation matrix. In a correlation matrix, the data have been standardized, with the variances set to 1.0. In a covariance matrix, the data entries are *unstandardized;* thus, it contains the same information, in unstandardized form, as a correlation matrix. The diagonal elements of a covariance matrix are variances—covariances of items with themselves—just as the unities along the main diagonal of a correlation matrix are variables' variances standardized to 1.0 and also their correlations with themselves. Its off-diagonal values are *covariances,* expressing relationships between pairs of unstandardized variables just as correlation coefficients do with standardization. So, conceptually, a covariance matrix consists of (a) variances (on the diagonal) for individual variables and (b) covariances (off the diagonal) representing the unstandardized relationship between variable pairs.

A typical covariance matrix for three variables (X_1, X_2, and X_3) is shown in Table 3.1.

	X_1	X_2	X_3
X_1	Var_1	$Cov_{1,2}$	$Cov_{1,3}$
X_2	$Cov_{1,2}$	Var_2	$Cov_{2,3}$
X_3	$Cov_{1,3}$	$Cov_{2,3}$	Var_3

Table 3.1 Variances and Covariances for Three Variables

The same matrix is displayed somewhat more compactly using the customary symbols for matrices, variances, and covariances:

$$\begin{bmatrix} \sigma_1^2 & \sigma_{1,2} & \sigma_{1,3} \\ \sigma_{1,2} & \sigma_2^2 & \sigma_{2,3} \\ \sigma_{1,3} & \sigma_{2,3} & \sigma_3^2 \end{bmatrix}$$

Covariance Matrices for Multi-Item Scales

Let us focus our attention on the properties of a covariance matrix for a set of items that, when added together, make up a scale. The covariance matrix presented above has three variables—X_1, X_2, and X_3. Assume that these variables are actually scores for three items and that the items (X_1, X_2, and X_3) when added together make up a scale we will call Y. What can this matrix tell us about the relationship of the individual items to the scale as a whole?

A covariance matrix has a number of interesting (well, useful, at least) properties. Among these is the fact that adding all the elements in the matrix together (i.e., summing the variances, which are along the diagonal, and the covariances, which are off the diagonal) gives a value that is exactly equal to the variance of the scale as a whole, assuming that the items are equally weighted. So, if we add all the terms in the symbolic covariance matrix, the resulting sum will be the variance of scale Y. This is very important and bears repeating: The variance of a scale (Y) made up of any number of items equals the sum of all the values in the covariance matrix for those items, assuming equal item weighting.[1] Thus, the variance of a scale (Y) made up of three equally weighted items (X_1, X_2, and X_3) has the following relationship to the covariance matrix of the items: $\sigma_y^2 = C$, where

$$C = \begin{bmatrix} \sigma_1^2 & \sigma_{1,2} & \sigma_{1,3} \\ \sigma_{1,2} & \sigma_2^2 & \sigma_{2,3} \\ \sigma_{1,3} & \sigma_{2,3} & \sigma_3^2 \end{bmatrix}.$$

Readers who would like more information about the topics covered in this section are referred to Nunnally (1978) for covariance matrices and Namboodiri (1984) for an introduction to matrix algebra in statistics. The covariance matrix for the individual items has other useful information. Applications that can be derived from item covariance matrices are discussed by Bohrnstedt (1969).

Alpha and the Covariance Matrix

Alpha is defined as the proportion of a scale's total variance that is attributable to a common source, presumably the true score of a latent variable underlying the items. Thus, if we want to compute alpha, it would be useful to have a value for the scale's total variance and a value for the proportion that is "common" variance. The covariance matrix is just what we need in order to do this.

Recall the diagram we used in Chapter 2 to show how items related to their latent variable, as in Figure 3.1.

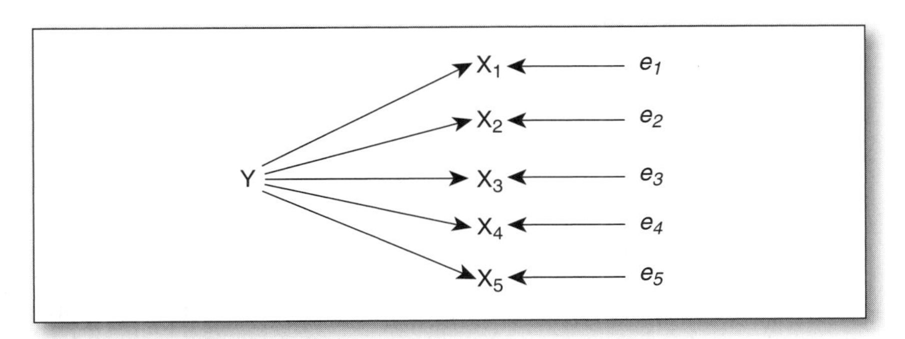

Figure 3.1 Diagrammatic representation of how a set of five items relates to the common latent variable Y

All the variation in items that is due to the latent variable, Y, is *shared* or *common.* (The terms *joint* and *communal* are also used to describe this variation.) When Y varies (as it will, for example, across individuals having different levels of the attribute it represents), scores on *all* the items will vary with it because it is a cause of those scores. Thus, if Y is high, all the item scores will tend to be high; if Y is low, they will tend to be low. This means that the items will tend to *vary jointly* (i.e., be correlated with one another). So the latent variable affects all the items and, thus, they are correlated. The error terms, in contrast, are the source of the *unique* variation that each item possesses. Whereas all items share variability due to Y, no two share any variation from the *same* error source under our classical measurement assumptions. The value of a given error term affects the score of only one item. Thus, the error terms are not correlated with one another. So, each item (and, by implication, the scale defined by the sum of the items) varies as a function of (a) the source of variation common to itself and the other items and (b) unique, unshared variation that we refer to as error. It follows that the total variance for each item—and, hence, for the scale as a whole—must be a combination of variance from common and unique sources. According to the definition of reliability, alpha should equal the ratio of common-source variation to total variation.

Now, consider a *k*-item measure called Y whose covariance matrix is as follows:

$$\begin{bmatrix} \sigma_1^2 & \sigma_{1,2} & \sigma_{1,3} & \cdots & \sigma_{1,k} \\ \sigma_{1,2} & \sigma_2^2 & \sigma_{2,3} & \cdots & \sigma_{2,k} \\ \sigma_{1,3} & \sigma_{2,3} & \sigma_3^2 & \cdots & \sigma_{3,k} \\ \vdots & \vdots & \vdots & \ddots & \vdots \\ \sigma_{1,k} & \sigma_{2,k} & \sigma_{3,k} & \cdots & \sigma_k^2 \end{bmatrix}$$

The variance (σ_y^2) of the *k*-item scale equals the sum of all matrix elements. The entries along the main diagonal are the variances of the individual items represented in the matrix. (The variance of the *i*th item is signified as σ_i^2.) Therefore, the sum of the elements along the main diagonal ($\Sigma\sigma_i^2$) is the sum of the variances of the individual items. Thus, the covariance matrix gives us ready access to two values: (a) the total variance of the scale (σ_y^2) defined as the sum of all elements in the matrix and (b) the sum of the individual item variances $\left(\Sigma\sigma_i^2\right)$ computed by summing entries along the main diagonal. These two values can be given a conceptual interpretation. The sum of the whole matrix is, by definition, the variance of Y, the scale made up of the individual items. However, this total variance, as we have said, can be partitioned into different parts.

Let us consider how the covariance matrix separates common from unique variance by examining how the elements on the main diagonal of the covariance matrix differ from all the off-diagonal elements. All the variances (diagonal elements) are single-variable or "variable-with-itself" terms. We noted earlier that these variances can be thought of as covariances of items with themselves. Each variance contains information about only one item. In other words, each represents information that is based on a single item, not joint variation shared among items. (Within that single item, some of its variation will be due to the common underlying variable and, thus, will be shared with other items; some will not. However, the item's variance does not quantify the extent of shared variance but merely the amount of dispersion in the scores for that item, irrespective of what causes it.) The off-diagonal elements of the covariance matrix all involve *pairs* of terms and, thus, common (or joint) variation between two of the scale's items (covariation). Thus, the elements in the covariance matrix (and, hence, the total variance of Y) consist of covariation (joint variation, if you will) plus "nonjoint" or "noncommunal" variation concerning items considered individually. Figure 3.2 pictorially represents these two subdivisions of the covariance matrix. The shaded area along the diagonal is the noncommunal portion of the matrix, and the two off-diagonal regions within the triangular borders are, together, the communal portion.

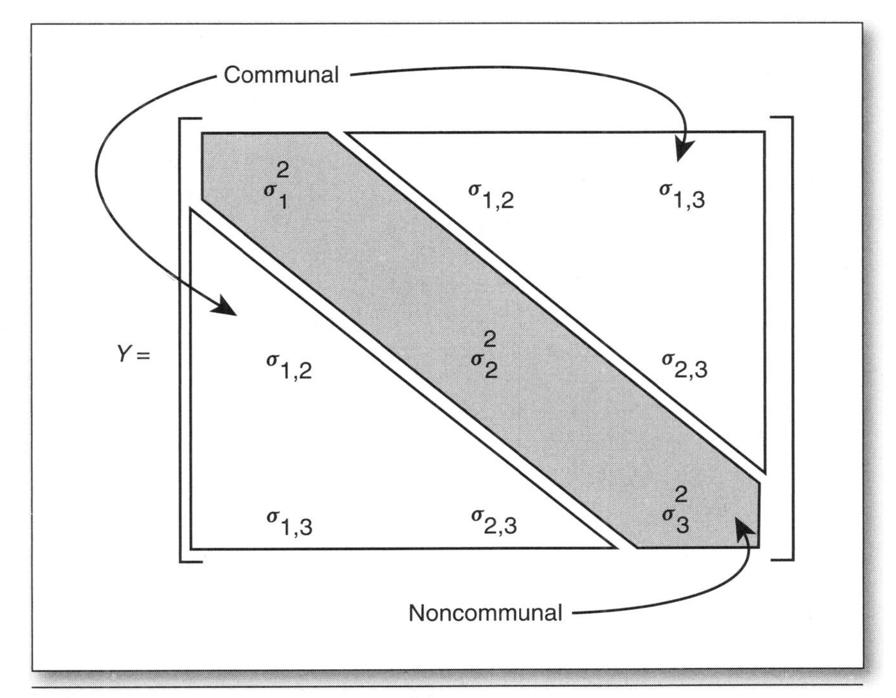

Figure 3.2 A variance-covariance matrix showing that the variances along the main diagonal are noncommunal, whereas the covariances lying above and below the diagonal are communal

As the covariances—and only the covariances—represent communal variation, all noncommunal variation must be represented in the variances along the main diagonal of the covariance matrix and, thus, by the term $\Sigma\sigma_i^2$. The total variance, of course, is expressed by σ_y^2, the sum of all the matrix elements. Thus, we can express the ratio of nonjoint variation to total variation in Y as

$$\sum\sigma_i^2/\sigma_y^2$$

This ratio corresponds to the sum of the diagonal values in the covariance matrix. It thus follows that we can express the proportion of joint, or communal, variation as what is left over—in other words, the complement of this value as shown:

$$1 - \left(\sum\sigma_i^2/\sigma_y^2\right)$$

This value corresponds to the sum of all the off-diagonal values of the covariance matrix. It may seem strange, or at least inefficient, to compute the diagonal elements and then subtract them from the value of the covariance matrix as a whole. Why not just compute the sum of the off-diagonal elements directly as $\Sigma\sigma_{i,j}$, where i and j represent each of the two items involved in a particular covariance? In fact, one would arrive at the same exact point by directly computing the sum of off-diagonal elements. The formula involving subtraction from 1 is a legacy of the days when computers were not available to do calculations. Computing the total variance for Y and the variance for each individual item (i) were probably operations that had already been done for other purposes. Even if there were no need to calculate these variances for other purposes, consider the computational effort involved. For a 20-item scale, the choice would be between computing 21 variances (one for each item and another for the entire scale) versus 190 covariances (i.e., one for each of the 380 off-diagonal elements of the matrix, with those above the diagonal identical to those below) plus the total variance. Thus, a formula that quantifies communal variance as what remains after removing noncommunal from total variance makes more practical sense than it might first appear to.

The value represented by the formula

$$1 - \left(\sum\sigma_i^2/\sigma_y^2\right)$$

or, equivalently,

$$\sum\sigma_{i,j}/\sigma_y^2$$

at first blush would seem to capture the definition of alpha (i.e., the communal portion of total variance in a scale that can be attributed to the items' common source, which we presume reflects the true score of the latent variable). We still need one more correction, however. This need becomes apparent if we consider what would happen if we had, say, five perfectly correlated items. Such an arrangement should result in perfect reliability. A correlation matrix, as noted earlier, is merely a covariance matrix in which all terms have been standardized. Because the terms in a correlation matrix are more familiar and consistent (i.e., a perfect association always corresponds to a correlation coefficient of 1.0), I will use a correlation matrix to make this point. If we used the expressions above to represent the portion of communal and total variance in our hypothetical correlation matrix of five perfectly reliable items, what would we get? The correlation matrix in this instance would consist of a 5×5 matrix with all values equal to 1.0. The denominator of the preceding equation, representing the total variance of the scale comprising the five items, would thus equal 25. The numerator, however, would equal only 20, thus yielding a reliability of 20/25 (or .80) rather than 1.0. Why is this so? The total number of elements in the covariance matrix is k^2. The number of elements in the matrix that are noncommunal (i.e., those along the main diagonal) is k. The number that are communal (all those *not* on the diagonal) is $k^2 - k$. Thus, the fraction in our last formula has a numerator based on $k^2 - k$ values and a denominator based on k^2 values. To adjust our calculations so that the ratio expresses the relative magnitudes rather than the numbers of terms that are summed in the numerator and denominator, we multiply the entire expression representing the proportion of communal variation by values to counteract the differences in numbers of terms summed. To do this, we multiply by $k^2 / (k^2 - k)$, or, equivalently, $k / (k - 1)$. This limits the range of possible values for alpha to between 0.0 and 1.0. In the five-item example just discussed, multiplying .80 by 5/4 yields the appropriate 1.0. Readers may want to do the mental arithmetic for matrices of other sizes. It should soon become apparent that $k / (k - 1)$ is always the multiplier that will yield an alpha of 1.0 when the items are all perfectly correlated. Thus, we arrive at the usual formula for coefficient alpha:

$$\alpha = \frac{k}{k-1}\left(1 - \frac{\sum \sigma_i^2}{\sigma_{y_i}^2}\right)$$

To summarize, a measure's reliability equals the proportion of total variance among its items that is due to the latent variable and thus is communal. The formula for alpha expresses this by specifying the portion of total variance for the item set that is unique, subtracting this from 1 to determine the proportion that is communal, and multiplying by a correction factor to adjust for the number of elements contributing to the earlier computations.

Alternative Formula for Alpha

Another common formula for computing alpha is based on correlations rather than covariances. Actually, it uses $\bar{r}$, the *average inter-item correlation.* This formula is

$$\alpha = \frac{k\bar{r}}{1 + (k - 1)\bar{r}}$$

It follows logically from the covariance-based formula for alpha. Consider the covariance formula in conceptual terms:

$$\alpha = \frac{k}{k - 1}\left(1 - \frac{\text{Sum of item variances}}{\text{Sum of variances and covariances}}\right)$$

Note that the numerator and denominator in the term on the right are sums of individual values. However, the sum of these individual values is identical to the mean of the values multiplied by the number of values involved. (For example, k numbers that sum to 50 and k times the mean of those numbers both equal 50. To illustrate further, substitute 10 for k in the preceding sentence; the average of 10 values that sum to 50 has to be 5, and 10 times 5 equals 50, the same value as the original sum.) Therefore, the numerator of the term on the right must equal k times the average item variance ($\bar{v}$) and the denominator must equal k times the average variance plus $(k^2 - k)$—or, alternatively, $(k)(k - 1)$—times the average covariance ($\bar{c}$):

$$\alpha = \frac{k}{k - 1}\left(1 - \frac{k\bar{v}}{k\bar{v} + (k)(k - 1)\bar{c}}\right)$$

To remove the 1 from the equation, we can replace it with its equivalent, $[k\bar{v} + (k)(k - 1)c]/[k\bar{v} + (k)(k - 1)c]$, which allows us to consolidate the whole term on the right into a single ratio:

$$\alpha = \frac{k}{k - 1}\left(\frac{k\bar{v} + k(k - 1)\bar{c} - k\bar{v}}{k\bar{v} + (k)(k - 1)\bar{c}}\right)$$

or, equivalently,

$$\alpha = \frac{k}{k - 1}\left(\frac{k(k - 1)\bar{c}}{k[\bar{v} + (k - 1)\bar{c}]}\right)$$

Cross-canceling k from the numerator of the left term and denominator of the right term while cross-canceling $(k - 1)$ from the numerator of the right term and denominator of the left term yields the simplified expression

$$\alpha = \frac{k\bar{c}}{\bar{v} + (k - 1)\bar{c}}$$

Recall that the formula we are striving for involves correlations rather than covariances and, thus, standardized rather than unstandardized terms. After standardizing, an average of covariances is identical to an average of correlations and a variance equals 1.0. Consequently, we can replace $\bar{c}$ with the average inter-item correlation $(\bar{r})$ and $\bar{v}$ with 1.0. This yields the correlation-based formula for coefficient alpha:

$$\alpha = \frac{k\bar{r}}{1 + (k - 1)\bar{r}}$$

This formula is known as the *Spearman-Brown prophecy formula*, and one of its important uses will be illustrated in the section of this chapter dealing with split-half reliability computation. In anticipation of that discussion, note that the formula presents reliability as a function of two scale properties: the number of items in the scale (k) and the average correlation between pairs of items $(\bar{r})$. Thus, the more items a scale contains and the stronger the average correlation among those items, the higher the reliability.

The two different formulas, one based on covariances and the other on correlations, are sometimes referred to as the *raw score* and *standardized score* formulas for alpha, respectively. The raw score formula preserves information about item means and variances in the computation process because covariances are based on values that retain the original scaling of the raw data. If items have markedly different variances, those with larger variances will have greater weight than those with lesser variances when this formula is used to compute alpha. The standardized formula based on correlations does not retain the original scaling metric of the items. Recall that a correlation is a standardized covariance. So all items are placed on a common metric and, thus, weighted equally in the computation of alpha by the standardized formula. Which is better depends on the specific context and whether equal weighting is desired. As we shall see in later chapters, recommended procedures for developing items often entail structuring their wording so as to yield comparable variances for each item. When these procedures are followed, there is typically little difference in the alpha coefficients computed by the two alternative methods. On the other hand, when procedures aimed at producing equivalent item variances are not followed, observing that the standardized and raw alpha values differ appreciably (e.g., by .05 or more) is indicative of at least one item having a variance that differs appreciably from the variances of the other items.

Critique of Alpha

One can argue that Cronbach's alpha is not an ideal indicator of reliability (e.g., Sijtsma, 2009). In fact, it is usually a lower bound for the actual reliability of a set of items rather than the best estimate of the actual reliability. Sijtsma (2009) suggests that the notion of internal consistency is somewhat ambiguous and questions to what extent alpha represents the most important aspects of internal consistency. He suggests that internal consistency is about the factor structure of a set of items and that regarding alpha as indicative of this is an oversimplification. In a strict sense, Sijtsma is correct; however, he acknowledges that there are problems with other estimates (e.g., a greatest-lower-bound estimate of reliability, or *glb*) that might be considered an alternative to alpha. He suggests, provocatively and appropriately, that as a pathway to greater precision, more work should be done to fully understand the types of bias that may exist in various indicators of reliability.

In my view, alpha retains utility. As a lower bound, it is a conservative estimate of reliability. Moreover, when care is taken in developing and selecting items and the items conform to unidimensionality, many of the potential pitfalls of using coefficient alpha as an indicator of reliability are abated. Finally, Cronbach's alpha has been widely used and researchers' norms about when instruments are sufficiently reliable largely have been based on the use of alpha. If alpha is perhaps a bit too conservative in some circumstances, then for any given level of reliability, some alternative indicator well might yield a higher numerical value than alpha. This might be confusing. Any given scale on any particular occasion will have a specific reliability. A method of estimating that reliability that yields a higher number does not change the reliability of the scale; it merely expresses it differently. If a given scale on a given occasion achieves a value for alpha of .80 and a value for some estimate of the glb of .82, the scale's reliability has not changed as a result of the latter calculation. Adopting a new standard for representing reliability would thus require a recalibration of our standards. This would not be a simple linear transformation; one cannot convert between a glb value and coefficient alpha by simply adding or multiplying by some fixed amount. (In the following chapter, I will discuss similar arguments against correcting correlations for attenuation when examining validity.)

Coefficient alpha has a strong conceptual linkage to the definition of reliability and to other indicators of reliability. Thus, to the extent that coefficient alpha has limitations, so do other indicators to which it is closely related. Alpha is a special case of the intraclass correlation coefficient (ICC), mentioned earlier. Both partition the variance observed into "true" and "error" components, and the ICC resulting from a situation like the objects placed in sun or shade, as in the earlier example, will yield an identical result to computing coefficient alpha. As we shall see later in this chapter, alpha is also closely related to other methods of reliability estimation, such as split-half reliability. Consequently, logic requires that abandoning these other methods of reliability assessment should accompany any rejection of coefficient alpha as an appropriate indicator of reliability.

As work continues on these alternatives to alpha, we well might reach a point in time when the advantages of these alternatives, relative to coefficient alpha, justify their wider adoption. In my judgment, for most researchers, we have not yet reached that point.

RELIABILITY BASED ON CORRELATIONS BETWEEN SCALE SCORES

There are alternatives to coefficient alpha as an indicator of reliability. These types of reliability computation may involve having the same set of people complete two separate versions of a scale or the same version on multiple occasions.

Alternate-Forms Reliability

If two strictly parallel forms of a scale exist, then the correlation between them can be computed as long as the same people complete both parallel forms. For example, assume that a researcher first developed two equivalent sets of items measuring patients' desire for control when interacting with physicians, then administered both sets of items to a group of patients, and then, finally, correlated the scores from one set of items with the scores from the other set. This correlation would be the alternate-forms reliability. Recall that parallel forms are made up of items, all of which (either within or between forms) do an equally good job of measuring the latent variable. This implies that both forms of the scale have identical alphas, means, and variances and measure the same phenomenon. In essence, parallel forms consist of one set of items that have more or less arbitrarily been divided into two subsets that make up the two parallel, alternate forms of the scale. Under these conditions, the correlation between one form and the other is equivalent to correlating either form with itself, as each alternate form is equivalent to the other.

Split-Half Reliability

A problem with alternate-forms reliability is that we usually do not have two versions of a scale that conform strictly to the assumptions of parallel tests. However, there are other reliability estimates that apply the same sort of logic to a single set of items. Because alternate forms are essentially made up of a single pool of items that have been divided in two, it follows that we can (a) take the set of items that make up a single scale (i.e., a scale that does not have any alternate form), (b) divide that set of items into two subsets, and (c) correlate the subsets to assess reliability.

A reliability measure of this type is called a split-half reliability. Split-half reliability is really a class rather than a single type of computational method because there are a variety of ways in which the scale can be split in half. One method is to compare the first half of the items with the second half. This type of *first-half, last-half split* may be problematic, however, because factors other than the value of the latent variable (in other words, sources of error) might affect each subset differently. For example, if the items making up the scale in question were scattered throughout a lengthy questionnaire, the respondents might be more fatigued when completing the second half of the scale. *Fatigue* would then differ systematically between the two halves and would thus make them appear less similar. However, the dissimilarity would not be so much a characteristic of the items per se as of their position in the item order of the scale. Other factors that might differentiate earlier-occurring from later-occurring items are a *practice effect* (whereby respondents might get better at answering the items as they go along), *failure to complete* the entire set of items, or possibly even something as mundane as changes in the *print quality* of a questionnaire from front to back. As with fatigue, these factors would lower the correlation between halves because of the order in which the scale items were presented and not because of the quality of the scale items. As a result of factors such as these, measuring the strength of the relationships among items may be complicated by circumstances not directly related to item quality, resulting in an erroneous reliability assessment.

To avoid some of the pitfalls associated with item order, one can assess another type of split-half reliability known as *odd-even reliability*. In this instance, the subset of odd-numbered items is compared with the even-numbered items. This assures that each of the two subsets of items consists of an equal number from each section (i.e., beginning, middle, end) of the original scale. Assuming that item order is irrelevant (as opposed to the "easy-to-hard" order common to achievement tests, for example), this method avoids many of the problems associated with first-half versus second-half splits.

In theory, there are many other ways to arrive at split-half reliability. Two alternatives to the methods discussed above for constituting the item subsets are *balanced halves* and *random halves*. In the former case, one would identify some potentially important item characteristics (e.g., first-person wording, item length, or whether a certain type of response indicates the presence or absence of the attribute in question). The two halves of the scale would then be constituted so as to have the characteristics equally represented in each half.

Thus, an investigator might divide the items so that each subset had the same number of items worded in the first person, the same number of short items, and so on. However, when considering multiple-item characteristics, it might be impossible to balance the proportion of one without making it impossible to balance another. This would be the case, for example, if there were more long than short first-person items. Creating a balance for the latter characteristic would necessitate an imbalance of the former. Also, it may be difficult to determine which characteristics of the items should be balanced.

An investigator could obtain random halves merely by randomly allocating each item to one of the two subsets that will eventually be correlated with each other to compute the reliability estimate. How well this works depends on the number of items, the number of characteristics of concern, and the degree of independence among the characteristics. Hoping that a small number of items, varying along several interrelated dimensions, will yield comparable groupings through randomization is unrealistic. On the other hand, randomly assigning a set of 50 items that vary with respect to two or three uncorrelated characteristics to two categories might yield reasonably comparable subsets.

Which method of achieving split halves is best depends on the particular situation. What is most important is that the investigator think about how dividing the items might result in nonequivalent subsets and what steps can be taken to avoid this. The reasoning behind both split-half and alternate-forms reliability is a natural extension of the parallel tests model.

Although when we initially discussed that model, we regarded each item as a test, one can also regard a scale (or the two halves of a scale) that conforms to the model as a test. Therefore, we can apply the logic we used in the case of several items to the case of two alternate forms or two halves of a scale. Consider two tests (scale halves or alternate forms) under the parallel tests assumptions, shown in Figure 3.3.

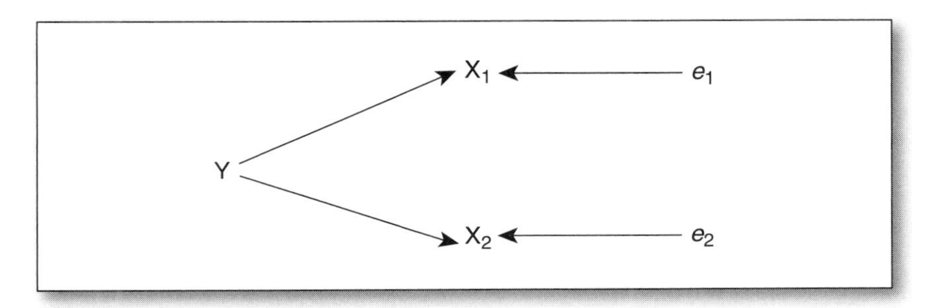

Figure 3.3 A path diagram showing the relationship of two split halves of a measure (X_1 and X_2) to their common latent variable

The only route linking the two consists of the causal paths from the latent variable to each split half. Thus, the product of these paths' values equals the correlation between the tests. If the path values have to be equal (and they do under the assumptions of this model), then the correlation between the tests equals the square of the path value from latent variable to either test. The square of that path (assuming it is a standardized path coefficient) is also the proportion of variance in either test that is influenced by the latent variable. This, in turn, is the definition of reliability. Thus, the correlation between the two tests equals the reliability of each.

Whereas the tests referred to in the preceding paragraph are two complete versions of a scale in the alternate-forms case, they are two half-scales in the split-half instance. Thus, the correlation between two split halves yields a reliability estimate for each *half* of the whole set of items, which is an underestimate of the reliability for the entire set. An estimate of the reliability of the entire scale, based on the reliability of a portion of the scale, can be computed by using the Spearman-Brown formula, discussed earlier in this chapter. Recall that, according to this formula,

$$\alpha = \frac{k\bar{r}}{1 + (k - 1)\bar{r}}$$

where k is the number of items in question and $\bar{r}$ is the average correlation of one item with any other (i.e., the average inter-item correlation). If you had determined the reliability of a subset of items (e.g., by means of the split-half method) and knew how many items that reliability was based on (e.g., half the number in the whole scale), you could use the formula to compute $\bar{r}$. Then, you could plug that value of $\bar{r}$ and the number of items in the *whole* scale back into the formula. The result would be an estimate of the reliability of the *whole* scale, based on a reliability value computed on split halves of the scale. It simplifies matters if you perform a little algebra on the Spearman-Brown equation to put it into the following form:

$$\bar{r} = \frac{r_{yy}}{[k - (k - 1)r_{yy}]}$$

where r_{yy} is the reliability of the item set in question. For example, if you knew that the split-half reliability for two nine-item halves was equal to .90, you could compute $\bar{r}$ as follows:

$$\bar{r} = \frac{.9}{[9 - (8)(.9)]} = .5$$

You could then recompute the reliability for the whole 18-item scale by using $r = .5$ and $k = 18$ in the Spearman-Brown formula. Thus, the reliability estimate for the full scale is

$$\frac{18 \times .5}{1 + (17 \times .5)}$$

which equals 9/9.5, or .947. Note that increasing the number of items has increased the reliability. A quick look at the Spearman-Brown formula should make it apparent that, all else being equal, a longer scale will always be more reliable than a shorter one. This may not be intuitively obvious. If each item has some signal and some noise, won't adding items increase the noise as well as the signal? The answer is that, yes, it will; however, Lord and Novick (2008) demonstrate that while error increases linearly, signal increases exponentially as items are added (p. 86)—that is, true score variance accumulates faster than error score variance as a test is lengthened. This is analogous to the way that any population parameter estimate (e.g., the estimated mean weight of adolescents in the United States) becomes progressively more precise (i.e., closer to the population true score) as the number of individuals sampled increases.

Inter-Rater Agreement

The reliability assessment methods we have discussed thus far have involved items as indicators. This is certainly appropriate for a book about scale development; however, in many research contexts, raters or judges serve as indicators. Although not strictly relevant to scale development, a brief discussion of inter-rater agreement (see DeVellis, 2005, for an overview of this topic) is useful because it can underscore the conceptual similarity it shares with item-based reliability assessment. In both cases, the underlying logic dictates that indicators sharing a common cause should be associated with one another. In the case of raters or judges, if the scores arising from their observations reflect properties of the observed stimulus rather than of the judges, then those scores should agree. The nature of that agreement can vary depending on the goals of the researcher and the approach taken to estimate inter-rater agreement.

In some cases, the investigator will require strict agreement in order to consider the ratings reliable. As an example, the extent to which two baseball umpires can reliably determine when a ball is fair or foul would depend on the extent to which they agreed over a series of occasions. Any hit ball must be either fair or foul according to baseball's rules. Thus, absolute agreement is appropriate as a means of gauging the reliability of the umpires' ratings. In other circumstances, raters may evaluate a phenomenon that varies along a continuum. Judges at a state fair who are rating the quality of locally baked pies, for example, are likely to use a scale that has more gradations than a simple good-bad dichotomy. In such a case, exact score agreement may not be necessary to determine that the ratings of multiple judges actually reflect an attribute of the pies and not just of the judges themselves. Although one judge may be more lenient than another and thus give consistently higher ratings, the scores across judges should roughly correspond if they are to be considered reliable. In fact, this might be a better basis for judging reliability in this circumstance than insisting on perfect agreement. Ratings assigned so that the order of pies scored from best to worst was similar across judges would suggest that it was the pies and not the judges that determined the assigned scores. Yet, if the judges' scoring criteria were calibrated somewhat differently (e.g., one judge considered a score of 5 to represent average pie quality, while another gave an average pie a score of 6), the judges might never actually agree. Thus, perfect agreement may not be an appropriate criterion for assessing reliability among judges in such a case.

In a highly influential paper, Patrick E. Shrout and Joseph L. Fleiss (1979) described the various ways of assessing inter-rater agreement under different circumstances. They emphasized that important features of the researcher's intent determined how best to assess inter-rater reliability. For the purposes of the present discussion, the most important feature they discussed was whether exact agreement or mere correspondence between judges best represented evidence of rater consistency. In the example of the umpires, clearly, absolute agreement would be necessary to conclude that the raters were reliable. The version of the ICC that is appropriate in this case is equivalent to Cohen's kappa (κ) coefficient (Cohen, 1960). This approach determines to what extent the frequency of exact agreements between judges exceeds what could be expected by chance. In contrast to the umpires, the pie judges could demonstrate that they were making reliable judgments without actually agreeing precisely. The form of the ICC that captures agreement under those circumstances is equivalent to Cronbach's coefficient alpha.

Thus, Cronbach's alpha can be used when one is interested in the extent to which either items or raters generate scores that correspond. In both cases, the correspondence (i.e., correlations among items or raters) arises because of the common influence of the underlying variable on the indicators, whether the latter are items or raters. The proportion of observed score variation that can be ascribed to the true score of the latent variable constitutes the indicators' reliability in both instances. This comparability of interpretation across different types of indicators (raters vs. items) underscores the general utility of alpha as an indicator of reliability.

Temporal Stability

Another method of computing reliability involves the temporal stability of a measure, or how constant scores remain from one occasion to another. *Test-retest reliability is* the method typically used to assess this. In some respects, test-retest reliability is analogous to split-half and alternate-forms reliability, discussed earlier. Returning to an earlier example, suppose that, instead of developing two sets of items to measure patients' desire for control when interacting with physicians, our hypothetical investigator developed a single-item set. Those items could be given to one group of patients on two separate occasions, and the scores from the first occasion could be correlated with those from the later administration. The rationale underlying reliability determinations of this type is that if a measure truly reflects some meaningful construct, it should assess that construct comparably on separate occasions. In other words, the true score of the latent variable should exert comparable influence on observed scores on two (or more) occasions, while the error component should not remain constant across administrations of the scale. Consequently, the correlation of scores obtained across two administrations of a scale to the same individuals should represent the extent to which the latent variable determines observed scores. This is equivalent to the definition of reliability as the proportion of variance attributable to the true score of the latent variable.

The problem with this reasoning is that what happens to the scores over time may or may not have to do with the error-proneness of the measurement procedure. Nunnally (1978) points out that items' characteristics might cause them to yield temporally stable responses even when the construct of interest has changed. For example, if a purported anxiety measure was influenced by social desirability as well as anxiety, scores might remain constant despite variations in anxiety. The stability in scores, reflected in a high correlation across occasions of administration, would not be the result of invariance in the phenomenon of interest. Alternatively, the phenomenon may not change while scores on the measure do—that is, the scale could be unreliable. Or changes in scores may be attributed to unreliability when, in fact, the phenomenon itself has changed and the measure has accurately tracked that change. The problem is that either a change or the absence of a change can be due to a variety of things besides the (un)reliability of the measurement procedure. Kelly and McGrath (1988) identified four factors that are confounded when one examines two sets of scores on the same measure, separated in time. These are (1) real change in the construct of interest (e.g., a net increase in average level of anxiety among a sample of individuals), (2) systematic oscillations in the phenomenon (e.g., variations in anxiety around some constant mean as a function of time of day), (3) changes attributable to differences in subjects or measurement methods rather than the phenomenon of interest (e.g., fatigue effects that cause items to be misread), and (4) temporal instability due to the inherent unreliability of the measurement procedure. Only the fourth factor is unreliability.

More recently, Yu (2005) has discussed how errors in test-retest reliability scores can arise not only from the shortcomings of the instrument itself but also from the individuals measured and the administration procedure. Errors arising from the participants being assessed include carryover effects in which performance on the first test influences subsequent performance. For example, a respondent may remember his or her earlier answers and may wish to appear consistent. Thus, the answers on the second administration are determined not directly by the state of the variable of interest but by a motivation to appear consistent over time. Yu refers to another respondent-based source of error—the deliberative effect. As an example, high school students may initially take the Scholastic Aptitude Test primarily as a way of gaining experience with the examination process. They may then deliberate on that experience and use it as a basis for preparing for a subsequent examination. As an alternative example of a deliberative effect, participants who find questions on an initial administration of some instrument offensive may intentionally omit or distort their responses when the items are readministered. Errors of administration mentioned by Yu include variations in procedures from test to retest (e.g., allowing respondents to go back and check earlier answers), poor instructions, subjectivity of the scoring of responses (e.g., criteria for scoring ambiguous answers), and an inappropriate time interval between administrations. Any of these types of errors could diminish reliability scores without the instrument itself being inherently unreliable.

This is not to say that demonstrating temporal stability is unimportant. In any number of research contexts, it may be critical to assume (or demonstrate) that measurements separated in time are highly correlated. However, the stability we seek in these situations is stability of both the *measure* and the *phenomenon.* Test-retest correlations tell us about the measure only when we are highly confident that the phenomenon has remained stable. Such confidence is not often warranted. Thus, test-retest reliability, although important, may best be thought of as revealing something about the nature of a phenomenon *and* its measurement, not the latter alone. Referring to invariance in scores over time as *temporal stability* is preferable because it does not suggest, as does test-retest reliability, that measurement error is the source of any instability we observe.

Computing test-retest reliability typically involves some form of the correlation coefficient. Two tests separated by time are analogous to two items administered at the same time. In the latter case, as noted earlier, the correlation between the items reflects the product of the two pathways extending from the latent variable to the items. Under the assumptions of strictly parallel tests, those pathways have identical values and so their product is equal to the squared value of each. Each pathway also represents the correlation between the latent variable at its originating end and the indicator at its terminating end. The square of that value is, thus, the squared correlation, or the proportion of variance shared between the latent variable and the indicator. So, by correlating the two items, we arrive at a value that is conceptually equivalent (under strictly parallel test assumptions) to the proportion of variance shared between the item and the latent variable. In the case of test-retest reliability, the logic is identical but the two indicators are now the instrument administered at Time 1 and then again at Time 2. Their correlation should estimate the squared correlation (and, thus, proportion of shared variance) between the obtained score (from either administration) and the true score of the latent variable. When the assumptions of strictly parallel tests are not met, the resultant value becomes a lower bound for the true reliability.

This analysis suggests that the reliability of a set of items administered on two occasions can be estimated by the correlation between the scores from the two administrations. This, in practice, is how test-retest reliability is typically computed. In the common circumstance when the scores on the instrument are essentially continuous, a Pearson product-moment correlation is typically used. The Pearson correlation should work well in the majority of instances for instruments developed using the approach described in this volume. Should the investigator be concerned that the scores in question do not conform to interval-level scaling (e.g., rankings), an alternative form of the correlation coefficient may be substituted. Note, however, that even if the individual items are not strictly interval, a score based on summing such items will more closely approximate interval scaling.

RELIABILITY AND STATISTICAL POWER

An often overlooked benefit of more reliable scales is that they increase *statistical power* for a given sample size (or allow a smaller sample size to yield equivalent power), relative to less reliable measures. To have a specified degree of confidence in the ability to detect a difference of a given magnitude between two experimental groups, for example, one needs a particular size sample. The probability of detecting such a difference (i.e., the power of the statistical test) can be increased by increasing the sample size. In many applications, much the same effect can be obtained by improving the reliability of measurement. A reliable measure, like a larger sample, contributes relatively less error to the statistical analysis. Researchers might do well to weigh the relative advantages of increasing scale reliability versus sample size in research situations where both options are available.

The power gains from improving reliability depend on a number of factors, including the initial sample size, the probability level set for detecting a Type I error, the effect size (e.g., mean difference) that is considered significant, and the proportion of error variance that is attributable to measure unreliability rather than sample heterogeneity or other sources. A precise comparison between reliability enhancement and sample size increase requires that these factors be specified; however, the following examples illustrate the point. In a hypothetical research situation with the probability of a Type I error set at .01, a 10-point difference between two means regarded as important, and an error variance equal to 100, the sample size would have to be increased from 128 to 172 (34% increase) to raise the power of an F test from .80 to .90. Reducing the total error variance from 100 to 75 (25% decrease) would have essentially the same result without increasing the sample size. Substituting a highly reliable scale for a substantially poorer one might accomplish this. As another example, for $N = 50$, two scales with reliabilities of .38 that have a correlation ($r = .24$) barely achieving significance at $p < .10$ are significant at $p < .01$ if their reliabilities are increased to .90. If the reliabilities remained at .38, a sample more than twice as large would be needed for the correlation to reach $p < .01$. Lipsey (1990) provides a more comprehensive discussion of statistical power, including the effects of measurement reliability.

Reliability is enhanced by both the number of items and the average correlation between items (which arises from a stronger correlation between each item and the true score); either more items or better items can increase reliability. Either method of enhancing reliability can enhance statistical power. Again, this is analogous to how sampling can increase statistical power. Reducing sampling error, either by sampling "bigger" (i.e., collecting data from a larger sample) or sampling "smarter" (i.e., collecting data from individuals who are better exemplars of the population of interest), can increase statistical power. Likewise, bigger (more items) and smarter (stronger inter-item correlations) scales can reduce variability arising from measurement error and, thus, increase statistical power.

GENERALIZABILITY THEORY

Thus far, our discussion of reliability has focused on partitioning observed variation into the portion that is attributable to the true score of the latent variable and the remaining portion, which is error. This section briefly introduces a more general framework for partitioning variance among error and nonerror sources. It harkens back to our earlier discussion of ANOVA-based approaches to distinguishing among sources of variation.

Before we apply the idea of a finer partitioning of error variance to measurement, let us consider a more general research example in which multiple sources of variation are examined. Suppose that a researcher wanted to determine the effectiveness of a training program intended to increase professional productivity. Assume, furthermore, that the researcher administered the training program to a large sample of college professors and to a comparable sample of artists. The researcher also identified comparable groups of professors and artists who would not participate in the training program but who would take part in the same productivity assessment as the training program participants. Upon giving the study some thought, this researcher might have concluded that the observations of productivity would reflect the operation of three identifiable sources of systematic variation: (a) participant versus nonparticipant, (b) professor versus artist, and (c) the interaction of these effects. A reasonable analytic strategy in this situation would be to perform ANOVA on the productivity scores, treating each of these sources of variation as a dimension in the analysis. The investigator could then determine to what extent each source of variation contributed to the total variation in professional productivity. In essence, this analytic strategy would partition the total variance among observed productivity scores into several sources: training participation, profession, the interaction of these, and error. Error would represent all sources of variation other than those specified by the preceding factors.

Now, consider a hypothetical situation in which a researcher is developing a scale of desire for autonomy. The measure will be used in a study of elderly people, some of whom may have visual problems. Consequently, the investigator plans to administer the desire-for-autonomy measure orally to those people who would have difficulty reading and in written form to the remaining study participants.

If the researcher ignored mode of administration (written vs. oral) as a source of variation in test scores, he or she would be regarding each score obtained as due to the true level of the respondent's desire for autonomy plus some degree of error. The researcher could proceed to calculate reliability as discussed earlier. Note, however, that merely computing alpha on the scale scores without regard for the mode of administration would not differentiate the potential systematic error due to administration method from any other source of error.

Alternatively, it is possible for the researcher to acknowledge administration mode as a source of variation among scores, using an ANOVA approach. If the resulting analysis demonstrated that the difference between administration methods accounted for an inconsequential proportion of the total variation in scores, then the researcher could have greater confidence in the comparability of scores for individuals completing either the oral or written version. If, on the other hand, a significant amount of the total observed variation in scores were attributable to administration mode, then the researcher would know that any interpretation of scores should take this difference between modes into consideration.

Generalizability theory (e.g., Cronbach, Gleser, Nanda, & Rajaratnam, 1972) provides a framework for examining the extent to which one can assume equivalence of a measurement process across one or more dimensions. In the preceding example, the dimension in question was mode of administration. Each dimension of interest is a potential source of variation and is referred to as a *facet.* The example focuses on mode of administration as the only potential source of variation (other than individuals) across which the investigator wished to generalize. Therefore, this example involves a single facet.

In the parlance of generalizability theory, observations obtainable across all levels of a facet (e.g., with both oral and written administration of the scale) constitute a *universe of admissible observations.* The mean of these observations is referred to as the *universe score* and is analogous to the true score of classical test theory (Allen & Yen, 1979). A study aimed at determining to what extent scores are comparable across different levels of a facet is called a generalizability study, or *G-study.* The hypothetical study of desire for autonomy is an example of a G-study by virtue of its addressing the effects of different levels of the mode-of-administration facet.

The purpose of the G-study is to help the investigator determine the extent to which the facet does or does not limit generalizability. If a facet (e.g., mode of administration) explains a significant amount of the variance in observed scores, findings *do not* generalize across levels (e.g., oral vs. written administration) of that facet. The extent to which one can generalize across levels of the facet without misrepresenting the data is expressed as a *generalizability coefficient.* This is typically computed by forming a ratio from the appropriate mean squares resulting from the ANOVA performed as part of the G-study. Conceptually, the generalizability coefficient is the ratio of universe score variance to observed score variance and is thus analogous to the reliability coefficient (Allen & Yen, 1979). Note, however, that if a G-study yields a poor generalizability coefficient, the study's design points to a source of the problem (i.e., the facet examined). A reliability coefficient merely identifies the amount of error without attributing it to a specific source.

In some instances, choosing the appropriate ANOVA design, deciding which effects correspond to the facets of interest, and constructing the correct generalizability coefficient can be demanding. Just as with ANOVA in general, multiple dimensions and nested, crossed, and mixed effects can complicate a G-study. (See Myers, 1979, or Kirk, 1995, for general discussions of ANOVA designs.) Keeping the design of a G-study simple is advisable. It is also prudent to consult a source that explains in detail how to build the appropriate ANOVA model for a given type of G-study. Crocker and Algina (1986) describe the appropriate designs for several different one- and two-facet generalizability studies. This source also provides a good general introduction to generalizability theory.

SUMMARY

Scales are reliable to the extent that they consist of reliable items that share a common latent variable. Coefficient alpha corresponds closely to the classical definition of reliability as the proportion of variance in a scale that is attributable to the true score of the latent variable. Various methods for computing reliability have different utility in particular situations. For example, if one does not have access to parallel versions of a scale, computing alternate-forms reliability is impossible. A researcher who understands the advantages and disadvantages of alternative methods for computing reliability is in a better position to make informed judgments when designing a measurement study or evaluating a published report. Reliability is an issue that transcends different research concerns and contexts. Generalizability theory and the intraclass correlation coefficient approach to reliability assessment both exploit the capacity of analysis of variance to isolate and quantify separate sources of variation. Thus, research activities as seemingly disparate as a generalizability study, a psychometric evaluation of questionnaire items, and an assessment of inter-rater reliability are all grounded in a common definition of reliability as the proportion of variance attributable to the true score of the phenomenon of interest.

EXERCISES[2]

1. If a set of items has good internal consistency, what does that imply about the relationship of the items to their latent variable?

2. In this exercise,[3] assume that the following is a covariance matrix for a scale (Y) made up of three items (X_1, X_2, and X_3):

$$\begin{matrix} .2 & .5 & .4 \\ .5 & .0 & .6 \\ .4 & .6 & .8 \end{matrix}$$

a. What are the variances of X_1, X_2, and X_3?
b. What is the variance of Y?
c. What is coefficient alpha for scale Y?

3. Discuss the ways in which test-retest reliability confounds other factors with the actual scale properties.

4. How does the logic of alternate-forms reliability follow from the assumptions of parallel tests?

NOTES

1. For weighted items, covariances are multiplied by products and variances by squares of their corresponding item weights. See Nunnally (1978, pp. 154–156) for a more complete description.

2. Throughout the book, the solution for any exercise requiring a numeric answer will appear in a footnote.

3. Answers: (a) 1.2, 1.0, and 1.8 (which sum to 4.0); (b) 7.0 (the sum of all elements in the matrix); (c) $(3/2) \times [1 - (4.0/7.0)] = .64$.

4

Validity

Whereas *reliability* concerns how much a variable influences a set of items, *validity* concerns whether the variable is the underlying cause of item covariation. To the extent that a scale is reliable, variation in scale scores can be attributed to the true score of some phenomenon that exerts a causal influence over all the items. However, determining that a scale is reliable does not guarantee that the latent variable shared by the items is, in fact, the variable of interest to the scale developer. The adequacy of a scale as a measure of a *specific variable* (e.g., perceived psychological stress) is an issue of validity.

Some authors have assigned a broader meaning to validity. For example, Messick (1995) described six types of validity, one of which (consequential validity) concerned the impact on respondents of how their scores are used. Although Messick's views on validity raised some thought-provoking issues, his classification system has not been widely adopted. According to the more conventional interpretation, validity is inferred from the manner in which a scale was constructed, its ability to predict specific events, or its relationship to measures of other constructs. There are essentially three types of validity that correspond to these operations:

1. Content validity
2. Criterion-related validity
3. Construct validity

Each type will be reviewed briefly. For a more extensive treatment of validity, including a discussion of methodological and statistical issues in criterion-related validity and alternative validity indices, see Chapter 10 in Ghiselli, Campbell, and Zedeck (1981). Readers might also want to consider Messick's (1995) more all-encompassing view of validity.

CONTENT VALIDITY

Content validity concerns item sampling adequacy—that is, the extent to which a specific set of items reflects a content domain. Content validity is easiest to evaluate when the domain (e.g., all the vocabulary words taught to sixth-graders) is well defined. The issue is more subtle when measuring attributes such as beliefs, attitudes, or dispositions because it is difficult to determine exactly what the range of potential items is and when a sample of items is representative. In theory, a scale has content validity when its items are a randomly chosen subset of the universe of appropriate items. In the vocabulary test example used above, this is easily accomplished. All the words taught during the school year would be defined as the universe of items. Some subset could then be sampled. However, in the case of measuring beliefs, for example, we do not have a convenient listing of the relevant universe of items. Still, one's methods in developing a scale (e.g., having items reviewed by experts for relevance to the domain of interest, as suggested in Chapter 5) can help maximize item appropriateness. For example, if a researcher needed to develop a measure contrasting expected outcomes and desired outcomes (e.g., expecting vs. wanting a physician to involve the patient in decision making), it might be desirable to establish that all relevant outcomes were represented in the items. To do this, the researcher might have colleagues familiar with the context of the research review an initial list of items and suggest content areas that have been omitted but that should be included. Items reflecting this content could then be added.

Content validity is intimately linked to the definition of the construct being examined. Stated simply, a scale's content should reflect the conceptual definition applicable *to that scale.* Some concepts may have been defined in more than one way by theoreticians or may lie at the intersection of multiple concepts. It is essential that item content capture the aspects of the phenomenon that are spelled out in its conceptual definition and not other aspects that might be related but are outside the investigator's intent for that particular instrument.

As an example, Sterba and colleagues (2007) set out to develop a measure of dyadic efficacy related to how couples in which one partner had rheumatoid arthritis perceived their ability to manage the illness as a team. This instrument was the first to assess *dyadic efficacy,* a couple's confidence in their ability *as a team* to take various health-promoting actions. Thus, the concept underlying the instrument was distinct from, although related to, other concepts such as individual self-efficacy. The first phase of this effort was an item development study that formed the basis for the authors' claims of content validity. This study aimed at identifying appropriate content from the broader empirical and theoretical literature for possible inclusion in the measure. Although the authors examined content from measures of related constructs (e.g., arthritis-specific self-efficacy), they geared their item development to specific features of the construct *as they had defined it.* A critical aspect of that definition involved how confident couples felt that they, *as a team,* could manage the challenges of the illness. Accordingly, the item development study included interviews with couples to get their thoughts about whether the construct rang true to them, to understand how they conceptualized it, and to identify the language they used to describe it. The insights gained from the conceptual definition of the construct, literature review, and patient interviews informed item construction. Items were written explicitly to capture the team aspect of couples' perceptions regarding efficacy. This process ensured that the item content reflected the *specific construct* in which the investigators were interested and not various other concepts (such as self-efficacy or perceived spousal support) that might be conceptually related to it. As an additional content validation step, Sterba et al. (2007) asked a group of content experts to review the items the research team initially developed vis-à-vis their conceptual definition. This procedure served as a further check that the items were representative of the relevant content the instrument was designed to measure. Collectively, these steps increased the likelihood that relevant content was included in the scale while irrelevant content was not, thus supporting claims of content validity.

CRITERION-RELATED VALIDITY

In order to have criterion-related validity, as the term implies, an item or scale is required only to have an empirical association with some criterion or putative "gold standard." Whether or not the theoretical basis for that association is understood is irrelevant to criterion-related validity. If one could show, for example, that dowsing is empirically associated with locating underground water sources, then dowsing would have validity with respect to the criterion of successful well digging. Thus criterion-related validity per se is more a practical issue than a scientific one, because it is not concerned with understanding a process but merely with predicting it. In fact, criterion-related validity is often referred to as *predictive validity.*

Criterion-related validity by any name does not necessarily imply a causal relationship among variables, even when the time ordering of the predictor and the criterion are unambiguous. Of course, prediction in the context of theory (e.g., prediction as a hypothesis) *may* be relevant to the causal relationships among variables and can serve a useful scientific purpose.

Another point worth noting about criterion-related validity is that, logically, one is dealing with the same type of validity issue whether the criterion follows, precedes, or coincides with the measurement in question. Thus, in addition to predictive validity, *concurrent validity* (e.g., predicting driving skill from answers to oral questions asked during the driving test) or even *postdictive validity* (e.g., predicting birth weight from an infancy developmental status scale) may be used more or less synonymously with criterion-related validity. The most important aspect of criterion-related validity is not the time relationship between the measure in question and the criterion whose value one is attempting to infer but, rather, the strength of the empirical relationship between the two events. The term *criterion-related validity* has the advantage over the other terms of being temporally neutral and, thus, is preferable.

Criterion-Related Validity Versus Accuracy

Before leaving criterion-related validity, a few words are in order concerning its relationship to accuracy. As Ghiselli et al. (1981) point out, the correlation coefficient, which has been the traditional index of criterion-related validity, may not be very useful when *predictive accuracy* is the issue. A correlation coefficient, for example, does not reveal how many cases are correctly classified by a predictor (although tables that provide an estimate of the proportion of cases falling into various percentile categories, based on the size of the correlation between predictor and criterion, are described by Ghiselli et al., p. 311). It may be more appropriate in some situations to divide both a predictor and its criterion into discrete categories and to assess the "hit rate" for placing cases into the correct category of the criterion based on their predictor category. For example, one could classify each variable into "low" versus "high" categories and conceptualize accuracy as the proportion of correct classifications (i.e., instances when the value of the predictor corresponds to the value of the criterion). Where one divides categories is an important consideration. Consider a criterion that has two nonarbitrary states, such as "sick" and "well," and an assessment tool that has a range of scores that an investigator wants to dichotomize. The purpose of the assessment tool is to predict whether people will test as positive or negative for the sickness in question. Because the outcome is dichotomous, it makes sense to make the predictor dichotomous. There are two possible errors in classification: The measure can mistakenly classify a truly sick person as well (false negative) or a truly well person as sick (false positive). Where along the range of scores from the assessment tool the dividing line is placed when dichotomizing can affect the rates of these two types of errors. At the extremes, classifying everyone as well will avoid any false negatives (but increase false positives) while classifying everyone as sick will avoid any false positives (but increase false negatives). Obviously, in both of these extreme cases, the assessment tool would have no predictive value at all. The goal, of course, is to choose a cutoff that produces the fewest errors of either type and, thus, the highest accuracy. Often, no ideal cut point (i.e., one resulting in perfect classification) exists. In such a case, the investigator may make a conscious effort to minimize one type of error rather than the other. For example, if the sickness is devastating and the treatment is effective, inexpensive, and benign, the cost of a false negative (resulting in undertreating) is far greater than that of a false positive (resulting in overtreating). Thus, choosing a cutoff so as to reduce false negatives while accepting false positives would seem appropriate. On the other hand, if the remedy is both expensive and unpleasant and the sickness mild, the opposite trade-off might make more sense.

Also, it is important to remember that, even if the correlation between a predictor measure and a criterion is perfect, the *score* obtained on the predictor is *not an estimate* of the criterion. Correlation coefficients are insensitive to linear transformations of one or both variables. A high correlation between two variables implies that scores on those variables obtained from the same individual will occupy similar locations on their respective distributions. For example, someone scoring very high on the first variable is likely also to score very high on the second if the two are strongly correlated. "Very high," however, is a relative rather than an absolute term and does not consider the two variables' units of measurement, for example. Transforming the predictor's units of measurement to that of the criterion may be necessary to obtain an accurate numerical prediction. This adjustment is equivalent to determining the appropriate intercept in addition to the slope of a regression line. A failure to recognize the need to transform a score could lead to erroneous conclusions. An error of this sort is perhaps most likely to occur if the predictor happens to be calibrated in units that fall into the same range as the criterion. Assume, for example, that someone devised the following "speeding ticket scale" to predict how many tickets drivers would receive over 5 years:

1. I exceed the speed limit when I drive.
   Frequently : Occasionally : Rarely : Never
2. On multilane roads, I drive in the passing lane.
   Frequently : Occasionally : Rarely : Never
3. I judge for myself what driving speed is appropriate.
   Frequently : Occasionally : Rarely : Never

Let us also make the implausible assumption that the scale correlates perfectly with the number of tickets received in a 5-year period. The scale is scored by giving each item a value of 3 when a respondent circles "frequently," 2 for "occasionally," 1 for "rarely," and 0 for "never." The item scores then are summed to get a scale score. The score's perfect criterion validity does not mean that a score of 9 translates into nine tickets over 5 years. Rather, it means that the people who score highest on the instrument are also the people who have the highest observed number of tickets in a year. Some empirically determined transformation (e.g., .33 × SCORE) would yield the actual estimate. This particular transformation would predict three tickets for a driver scoring 9. The higher the criterion-related validity, the more accurate the estimate based on the predictor measure can be. However, the similarity between the numerical values of the criterion and the predictor measure prior to an appropriate transformation would have nothing to do with the degree of validity.

CONSTRUCT VALIDITY

Construct validity (Cronbach & Meehl, 1955) is directly concerned with the theoretical relationship of a variable (e.g., a score on some scale) to other variables. It is the extent to which a measure "behaves" the way that the construct it purports to measure should behave with regard to established measures of other constructs. So, for example, if we view some variable, based on theory, as positively related to constructs *A* and *B*, negatively related to *C* and *D*, and unrelated to *X* and *Y*, then a scale that purports to measure that construct should bear a similar relationship to measures of those constructs. In other words, our measure should be positively correlated with measures of constructs *A* and *B*, negatively correlated with measures of *C* and *D*, and uncorrelated with measures of *X* and *Y*. A depiction of these hypothesized relationships might look like the one in Figure 4.1.

	A	B	C	D	X	Y
Variable	+	+	–	–	0	0

Figure 4.1 A hypothesized relationship among variables

The extent to which empirical correlations matched the predicted pattern provides some evidence of how well the measure behaves as does the variable it is supposed to measure.

Applying this general approach, Sterba et al. (2007) described a pattern of associations they predicted their instrument would have with several relevant constructs. For example, based on their theoretical analysis, they predicted that dyadic efficacy scores would correlate with measures of marital quality, psychological adjustment, and teamwork standards. Some of these predictions specified negative associations, such as a hypothesized inverse relationship between dyadic efficacy and depression. A limitation of that study, acknowledged by the authors, was that individual self-efficacy was not assessed. Demonstrating that dyadic efficacy is distinct from the individual self-efficacy perceptions of partners within couples is a potentially important aspect of the ongoing process of scale validation.

Differentiating Construct From Criterion-Related Validity

People often confuse construct and criterion-related validity because the same exact correlation can serve either purpose. The difference resides more in the investigator's intent than in the value obtained. For example, an epidemiologist might attempt to determine which of a variety of measures obtained in a survey study correlate with health status. The intent might be merely to identify risk factors without concern (at least initially) for the underlying causal mechanisms linking scores on measures to health status. Validity, in this case, is the degree to which the scales can predict health status. Alternatively, the concern could be more theoretical and explanatory. The investigator, like the epidemiologist described in this book's opening chapter, might endorse a theoretical model that views stress as a cause of health status, and the issue might be how well a newly developed scale measures stress. This might be assessed by evaluating the behavior of the scale relative to how theory suggests stress should operate. If the theory suggested that stress and health status should be correlated, then the same empirical relationship used as evidence of predictive validity in the preceding example might be used as evidence of construct validity.

So-called *known-groups validation* is another example of a procedure that can be classified either as construct or criterion-related validity, depending on the investigator's intent. Known-groups validation typically involves demonstrating that some scale can differentiate members of one group from one another based on their scale scores. The purpose may be either theory related (as when a measure of attitudes toward a certain group is validated by correctly differentiating those who do or do not affiliate with members of that group) or purely predictive (as when one uses a series of seemingly unrelated items to predict job turnover). In the former case, the procedure should be considered a type of construct validity and in the latter, criterion-related validity.

In addition to intent, another difference often underlies criterion and construct validity. Criterion validity is often assessed directly by computing a correlation between the measure being validated and the criterion (e.g., some behavior, status, or score). In contrast, construct validity can be assessed only indirectly (see Lord & Novick, 2008, p. 278), because the relevant comparison is to a latent rather than an observed variable. In this regard, it is similar to reliability. In previous chapters, we noted that reliability is really about the relationship of an indicator and an unobservable true score and that we infer that relationship on the basis of correlations among observable indicators. The same is true for construct validity. We cannot directly compute the association between the instrument being validated and the latent variable but must do so indirectly by observing the associations between the new instrument and other credible indicators of the latent variable. In fact, investigators may find it useful to use this difference as a rough rule of thumb when they are uncertain whether criterion or construct validity is of primary interest. If the objective is to predict an observable outcome (e.g., behavior, status, or an observed score), then criterion validity may well be the goal. On the other hand, if the objective is to predict the level of some hypothetical, unobservable construct by means of an observable indicator, the goal is likely to be construct validity.

Sometimes, however, criterion validation will involve an indicator intended to reflect some unobservable true score. This is the case when the criterion (e.g., a law school graduate qualifying for admission to the bar) is based on some form of psychometric assessment (e.g., passing the bar examination). If the investigator's goal remains sheer prediction without conceptual elaboration, such a comparison is an instance of criterion validity despite the indirectness of the comparison between the predictor and the outcome of interest.

Attenuation

To the extent that two indicators are not perfectly reliable, any correlation between them will underestimate the correlation between their corresponding true scores (e.g., Lord & Novick, 2008). When we assume that error is random, only the reliable portions of the two indicators can correlate. Accordingly, an observed validity coefficient computed by correlating two observed variables is attenuated as a consequence of any unreliability inherent to those indicators. One can apply a correction for attenuation to an observed correlation that takes into account the unreliability of the variables. Such a correction involves dividing the observed correlation between the two indicators by the square root of the product of their reliabilities. Thus,

$$r(T)_{xy} = \frac{r_{xy}}{\sqrt{r_{xx}r_{yy}}}$$

where $r(T)_{xy}$ is the correlation between the true scores for variables X and Y, r_{xy} is the correlation between the observed scores, and r_{xx} and r_{yy} are the reliabilities for variables X and Y, respectively.

Despite the availability of such corrections, some authors argue against them. Nunnally and Bernstein (1994), for example, point out that they may mislead investigators into thinking that an association is stronger than it actually is. They also note that corrected coefficients can sometimes exceed 1.0 (p. 257), which is problematic. Lord and Novick (2008) point out that when the reliabilities of the two measures (which occur in the denominator of the correction equation) are underestimated, the correlation between true scores will be overestimated (p. 138). Recalling from earlier chapters that under models other than parallel tests, coefficient alpha is a lower bound to the true reliability of a measure, it becomes clear that an inaccurate "correction" could easily arise. Thus, there are strong practical arguments against using a correction for attenuation when examining correlations, whether for assessing validity or for other purposes. Because of the indirect nature of the correlation coefficients used to assess construct validity, as discussed earlier, it may be especially tempting to correct those validity coefficients for attenuation. However, the arguments against adjusting for attenuation still apply, and it is common practice not to correct the correlation coefficient in such instances.

How Strong Should Correlations Be to Demonstrate Construct Validity?

There is no cutoff that defines construct validity. It is important to recognize that two measures may share more than construct similarity. Specifically, similarities in the way that constructs are measured may account for some covariation in scores independent of construct similarity. For example, two variables scored on a multipoint scoring system (scores from 1–100) will have a higher correlation with one another than with a binary variable, all else being equal. This is an artifact caused by the structure of the measurement methods. Likewise, because of procedural similarities, data of one type gathered by interviews may correlate to a degree with other data gathered in the same way—that is, some of the covariation between two variables may be due to measurement similarity rather than construct similarity. This fact provides some basis for answering the question concerning the magnitude of correlations necessary to conclude construct validity. The variables, at a minimum, should demonstrate covariation above and beyond what can be attributed to shared method variance.

Multitrait-Multimethod Matrix

Campbell and Fiske (1959) devised a procedure called the multitrait-multimethod matrix that is extremely useful for examining construct validity.

This approach fits well with the idea, presented earlier, that construct validity is assessed indirectly and must be inferred from available indicators of the latent variable of interest. The procedure involves measuring more than one construct by means of more than one method so that one obtains a "fully crossed" method-by-measure matrix. For example, suppose that a study is designed in which anxiety and depression and shoe size are each measured at two separate times using two different measurement procedures each time. (Note that two different samples of individuals could have been measured at the same time. What effect would this have on the logic of the approach?) Each construct could be assessed by two methods: a visual-analog scale (a line on which respondents make a mark to indicate the amount of the attribute they possess, be it anxiety, depression, or bigness of foot) and a rating assigned by an interviewer following a 15-minute interaction with each subject. One could then construct a matrix of correlations obtained between measurements as shown in Table 4.1.

Another possible distinction, not in the table, is between related versus *unrelated* traits. Because the entries that reflect the same trait (construct) and

Time 2 \ Time 1	A_v	A_i	D_v	D_i	S_v	S_i
A_v	**TM**	**T**	**M**		**M**	
A_i	**T**	**TM**		**M**		**M**
D_v	**M**		**TM**	**T**	**M**	
D_i		**M**	**T**	**TM**		**M**
S_v	**M**		**M**		**TM**	**T**
S_i		**M**		**M**	**T**	**TM**

Table 4.1 Interpretations of Correlations in a Multitrait-Multimethod Matrix

Notes:

TM = same trait and method (reliability); T = same trait, different method; M = same method, different trait.

A, *D*, and *S* refer to the constructs anxiety, depression, and shoe size, respectively. Subscripts *v* and *i* refer to visual-analog and interview methods, respectively.

the same method should share both method and construct variance, one would expect these correlations to be highest. It is hoped that correlations corresponding to the same trait but different methods would be the next highest. If so, this would suggest that construct covariation is higher than method covariation. In other words, our measures were more influenced by *what* was measured than by *how* it was measured. In contrast, there is no reason why any covariation should exist between shoe size and either of the other two constructs when they are measured by different procedures. Thus, these correlations should not be significantly different from zero. For nonidentical but theoretically related constructs, such as depression and anxiety, one would expect some construct covariation. This is potentially a highly informative set of correlations for establishing construct validity. If, for example, our depression measures were both well established but our anxiety measures were currently being developed, we could assess the amount of covariation attributable to concept similarity under conditions of similar and different measurement procedures. Theory asserts that anxiety and depression should be substantially correlated even when measured by different methods. If this proved to be the case, it would serve as evidence of the construct validity of our new anxiety measures. More specifically, these correlations would be indicative of *convergent validity,* evidence of similarity between measures of theoretically related constructs. Ideally, the correlations between anxiety and depression would be less than those between two depression or two anxiety measures but substantially greater than between either of the depression scores and shoe size. Equally important is evidence that the anxiety measures did not correlate significantly with measures of shoe size, irrespective of similarity or dissimilarity of measurement technique. This is evidence of *discriminant validity* (sometimes called divergent validity), the absence of correlation between measures of unrelated constructs. Shoe size and anxiety correlating significantly when measured the same way would suggest that method per se accounted for a substantial amount of the variation (and covariation) associated with similar measures of the dissimilar constructs.

Mitchell (1979) observed that the methods involved in collecting data for a multitrait-multimethod matrix constitute a two-facet G-study (or generalizability study; see Chapter 3), with traits and methods being the facets. The multitrait-multimethod matrix allows us to partition covariation into "method" and "trait" (or construct) sources. We can then make more precise statements about construct validity, because it allows us to differentiate covariation that truly reflects similarity of construct (and thus is relevant to construct validity) from covariation that is an artifact of applying similar measurement procedures (and thus does not relate to construct validity). Such a differentiation is not possible when one simply examines a single correlation between two measures.

WHAT ABOUT FACE VALIDITY?

Many people use the term *face validity* to describe a set of items that assess what they appear to measure on their face. In my view, this usage is unfortunate for several reasons.

First, the assumption that a measure assesses what it looks as though it is assessing can be wrong. For example, Idler and Benyamini (1997) examined 27 large, well-conducted epidemiological studies to determine precisely what a common item was tapping. That item asks people to evaluate their overall health as poor, fair, good, or excellent. Most people would judge this single-item measure to assess exactly what it says: the respondents' health. Idler and Benyamini noted that the item was an excellent predictor of a variety of health outcomes. It consistently outperformed other variables in accounting for variance across the different studies. More relevant to our discussion, it appeared not to be related primarily to health status. Models often contained the single item and also established measures of health status. Typically, both the single-item health self-rating and the other health status measures were significant predictors in the same model. That is, they did not share sufficient variance for the predictive contribution of one to preclude an independent predictive contribution from the other. Rather, the single-item health self-rating appeared to share variance to a greater degree with psychological variables. These findings suggest that this extensively used single item is *not* a valid indicator of health status, as it appears on its face. For this item, looking as though it is measuring what we want it to measure is not enough to support claims of validity.

A second problem with evaluating a measure based on face validity is that there are times when it is important that the variable being measured not be evident. For example, an instrument intended to assess the degree to which people answer untruthfully (e.g., to make themselves "look good") would hardly benefit from having its purpose apparent to respondents. Would we conclude that it was invalid because it did not look as if it were measuring untruthfulness? Hopefully not. So, here we have a case where failure to look like what it actually is cannot support a conclusion of invalidity.

A final concern with face validity is that it is unclear to whom an instrument's purpose should be evident, on its face. Is it the respondent? If a physician asks a patient if he or she has been more thirsty than usual, is the validity of that question dependent on the patient knowing what he or she was asked?

Clearly not. Is it the person creating the instrument who should recognize the purpose? It is hard to imagine that the linkage between instrument content and variable of interest is not obvious to an instrument's creator (except perhaps in cases concerning purely empirical, atheoretical, criterion-related validity). If this meaning of face validity were adopted, all scales essentially would be judged valid. Finally, is it a broader scientific community that should recognize an instrument's purpose based on its appearance? This interpretation is likely to yield conflicting evidence. An item that looks to some experts as though it measures one variable might look as though it measures another to a second, equally qualified group. Often, it seems people who claim that a scale is or is not valid because it does or does not appear to have face validity are basing their claim on personal perceptions. That is, if the intent and appearance of an instrument look similar *to them,* they are inclined to consider it face valid; otherwise, they are not. This seems like a feeble basis for any claim of validity.

Face validity and content validity are sometimes confused because both may concern the extent to which item content appears relevant to the construct of interest. An important difference, however, is that content validity is defined in terms of specific procedures, and those procedures are generally more structured and rigorous than informal assessments of face validity. As noted, these steps may include gathering insights from potential respondents but also typically include obtaining information from the relevant theoretical literature and from experts in the field under investigation. For example, content experts may be asked to evaluate item content in relation to an explicit construct definition. Such experts presumably have a theoretical frame of reference that supports their judgments. A mere appearance of relevance is not the sole criterion. Also, the individuals whose judgments are used to assess validity are clear. Moreover, concluding that the instrument has content validity is not solely the determination of the investigators themselves. Finally, transparency to the respondent is not a basis for evaluating content validity as it is typically assessed (e.g., by expert judgments). Thus, a formal approach to content validation can overcome the limitations of relying on face validity.

Depending on the circumstances, there may be advantages or disadvantages to an instrument's intent being evident from its appearance. As we shall see in the next chapter, the item generation process often produces statements that refer explicitly to the variable of interest. This usually is not a bad thing. I am not suggesting that instruments generally should be constructed so that their intent is not evident from appearances; rather, I am suggesting that whether or not that is the case has little or nothing to do with validity.

EXERCISES

1. Give an example of how the same correlation between a scale and a behavior might be indicative of either construct validity or criterion-related validity. Explain how (a) the motives behind computing the correlation and (b) the interpretation of that correlation would differ depending on the type of validity the investigator was trying to assess.

2. Assume that an investigator has paper-and-pencil measures of two constructs: self-esteem and social conformity. The investigator also has interview-based scores on the same two constructs. How could these data be used in a multitrait-multimethod matrix to demonstrate that the method of data collection had an undesirably strong effect on the results obtained?

5

Guidelines in Scale Development

Thus far, our discussion has been fairly abstract. We now look at how this knowledge can be applied. This chapter provides a set of specific guidelines that investigators can use in developing measurement scales.

STEP 1: DETERMINE CLEARLY WHAT IT IS YOU WANT TO MEASURE

This is deceptively obvious, and many researchers *think* they have a clear idea of what they wish to measure only to find that their ideas are vaguer than they thought. Frequently, this realization occurs after considerable effort has been invested in generating items and collecting data—a time when changes are far more costly than if discovered at the outset of the process. Should the scale be based in theory, or should you strike out in new intellectual directions? How specific should the measure be? Should some aspect of the phenomenon be emphasized more than others?

Theory as an Aid to Clarity

As noted in Chapter 1, thinking clearly about the content of a scale requires thinking clearly about the construct being measured. Although there are many technical aspects involved in developing and validating a scale, one should not overlook the importance of being well grounded in the substantive theories related to the phenomenon to be measured. The types of scales that are the primary focus of this book are intended to measure elusive phenomena that cannot be observed directly. Because there is no tangible criterion against which one can compare this type of scale's performance, it is important to have some clear ideas to serve as a guide. The boundaries of the phenomenon must be recognized so that the content of the scale does not inadvertently drift into unintended domains.

Theory is a great aid to clarity. Relevant social science theories should *always* be considered before developing a scale of the type discussed in this volume. If it turns out that extant theory offers no guide to the scale developers, then they may decide that a new intellectual direction is necessary. However, this decision should be an informed one, reached only after reviewing appropriate theory related to the measurement problem at hand. Even if there is no available theory to guide the investigators, they must lay out their own conceptual formulations prior to trying to operationalize them. In essence, they must specify at least a tentative theoretical model that will serve as a guide to scale development. This may be as simple as a well-formulated definition of the phenomenon they seek to measure. Better still would be to include a description of how the new construct relates to existing phenomena and their operationalizations.

Specificity as an Aid to Clarity

The level of specificity or generality at which a construct is measured also may be important. There is general agreement in the social sciences that variables will relate most strongly to one another when they match with respect to level of specificity (see Ajzen & Fishbein, 1980, for a discussion). Sometimes, a scale is intended to relate to very specific behaviors or constructs, while at other times, a more general and global measure is sought.

As an illustration of measures that differ in specificity, consider the locus of control construct. Locus of control is a widely used concept that concerns individuals' perceptions about who or what influences important outcomes in their lives. The construct can be applied broadly, as a means of explaining patterns of global behavior spanning many situations, or narrowly, to predict how an individual will respond in a very specific context. The sources of influence also can be described either broadly or specifically. Rotter's (1966) Internal-External scale, for example, is concerned at a fairly general level with these perceptions. A single dimension ranging from personal control to control by outside factors underlies the scale, and the outcomes on which the items focus are general, such as personal success. The external sources of control also are described in general terms. The following external statement is from Rotter's Internal-External scale: "The world is run by the few people in power, and there is not much the little guy can do about it."

Levenson (1973) developed a multidimensional locus of control scale that allows for three loci of control: oneself, powerful other people, and chance or fate. This permits an investigator to look at external sources of control a bit more specifically by characterizing them as either powerful others or fate. The outcomes on which she focused, however, remained general. An example of an item from Levenson's Powerful Others subscale is "I feel like what happens in my life is determined by powerful others."

Wallston, Wallston, and DeVellis (1978) developed the Multidimensional Health Locus of Control (MHLC) scales using Levenson's three loci of control, with outcomes specific to health, such as avoiding illness or getting sick. A sample item from the Powerful Others scale of the MHLC is "Having regular contact with my physician is the best way for me to avoid illness." Wallston, Stein, and Smith (1994) subsequently developed an even more outcome-specific health locus of control measure (MHLC Form C) that consists of a series of "template" items. This measure allows the researcher to specify any health problem of interest by substituting the name of the illness or disorder for the phrase "my condition" in each of the template items. A sample item from the Powerful Others scale of MHLC Form C, as it might be used in a study of diabetes, is "If I see my doctor regularly, I am less likely to have problems with my diabetes."

Each of these progressively more specific locus of control scales is potentially useful. Which is most useful depends largely on what level of outcome or locus generality relates to the scientific question being asked. For example, if a locus of control scale is intended to predict a general class of behavior or will be compared with other variables assessing constructs at a general level, then Rotter's scale may be the best choice because it, too, is general. On the other hand, if a researcher is interested in predicting specifically how beliefs about the influence of other people affect certain health behaviors, then the Wallston et al. (1994) scale may be more appropriate because the level of specificity matches that research question. During its development, each of these scales had a clear frame of reference that determined what level of specificity was appropriate, given the intended function of the scale. The point is that scale developers should make this determination as an active decision and not merely generate a set of items and then see what they look like after the fact.

The locus of control example illustrated specificity with respect to outcomes (e.g., how the world is run vs. problems with diabetes) and the loci of control (i.e., external generally vs. fate and powerful others separately). However, scale specificity can vary along a number of dimensions, including content domains (e.g., anxiety vs. psychological adjustment more broadly), setting (e.g., questionnaires designed specifically for relevance to particular work environments), or population (e.g., children vs. adults or military personnel vs. college students).

Being Clear About What to Include in a Measure

Scale developers should ask themselves if the construct they wish to measure is distinct from other constructs. As noted earlier, scales can be developed to be relatively broad or narrow with respect to the situations to which they apply. This is also the case with respect to the constructs they cover. Measuring general anxiety is perfectly legitimate. Such a measure might assess both test anxiety and social anxiety. This is fine if it matches the goals of the scale developer or user. However, if one is interested in only one specific type of anxiety, then the scale should exclude all others. Items that might "cross over" into a related construct (e.g., tapping social anxiety when the topic of interest is test anxiety) can be problematic.

Sometimes, apparently similar items may tap quite different constructs. In such cases, although the purpose of the scale may be to measure one phenomenon, it may also be sensitive to other phenomena. For example, certain depression measures, such as the Center for Epidemiological Studies Depression Scale (Radloff, 1977), have some items that tap somatic aspects of depression (e.g., concerning the respondent's ability to "get going"). In the context of some health conditions, such as arthritis, these items might mistake aspects of the illness for symptoms of depression (see Blalock, DeVellis, Brown, & Wallston, 1989, for a discussion of this specific point). A researcher developing a new depression scale might choose to avoid somatic items if the scale was to be used with certain populations (e.g., the chronically ill) or with other measures of somatic constructs (such as hypochondriasis). Used for other purposes, of course, it might be very important to include somatic items, as when the line of investigation specifically concerns somatic aspects of negative affect.

STEP 2: GENERATE AN ITEM POOL

Once the purpose of a scale has been clearly articulated, the developer is ready to begin constructing the instrument in earnest. The first step is to generate a large pool of items that are candidates for eventual inclusion in the scale.

Choose Items That Reflect the Scale's Purpose

Obviously, these items should be selected or created with the specific measurement goal in mind. The description of exactly what the scale is intended to do should guide this process. Recall that all items making up a homogeneous scale should reflect the latent variable underlying them. Each item can be thought of as a test, in its own right, of the strength of the latent variable. Therefore, the content of each item should primarily reflect the construct of interest. Multiple items will constitute a more reliable test than individual items, but each must still be sensitive to the true score of the latent variable.

Theoretically, a good set of items is chosen randomly from the universe of items relating to the construct of interest. The universe of items is assumed to be infinitely large, which pretty much precludes any hope of actually identifying it and extracting items randomly. However, this ideal should be kept in mind. If you are writing items anew, as is so often the case, you should think creatively about the construct you seek to measure. What other ways can an item be worded so as to get at the construct? Although the items should not venture beyond the bounds of the defining construct, they should exhaust the possibilities for types of items within those bounds. The properties of a scale are determined by the items that make it up. If they are a poor reflection of the concept you have worked long and hard to articulate, then the scale will not accurately capture the essence of the construct.

It is also important that the "thing" the items have in common be truly a construct and not merely a category. Recall, once again, that our models for scale development regard items as overt manifestations of a common latent variable that is their cause. Scores on items related to a common construct are determined by the true score of that construct. However, as noted in Chapter 1, just because items relate to a common category, that does not guarantee that they have the same underlying latent variable. Such terms as attitudes, barriers to compliance, or life events often define categories of constructs rather than the constructs themselves. A pool of items that will eventually be the basis of a unidimensional scale should not merely share a focus on attitudes, for example, but on *specific* attitudes, such as attitudes toward punishing drug abusers. One can presumably envision a characteristic of the person—a latent variable, if you will—that would "cause" responses to items dealing with punishing drug abusers. It is quite a challenge to imagine a characteristic that accounts for attitudes in general. The same is true for the other examples cited. Barriers to compliance are typically of many types. Each type (e.g., fear of discovering symptoms, concern over treatment costs, anticipation of pain, distance of treatment facilities, perceptions of invulnerability) may represent a latent variable. There may even be nontrivial correlations among some of the latent variables. However, each of these barriers is a separate construct. Thus, the term *barriers* describes a category of constructs rather than an individual construct related to a single latent variable. Items measuring different constructs that fall within the same category (e.g., perceptions of invulnerability and concerns over treatment costs) should not be expected to covary the way items do when they are manifestations of a common latent variable.

Redundancy

Paradoxically, redundancy is both a good and a bad feature of items within a scale. Resolving this paradox entails distinguishing between item features that strengthen a scale through repetition and those that do not. Because this topic is often a source of confusion, I will discuss it in some detail. I will first make the case in favor of redundancy.

At this stage of the scale development process, it is better to be more inclusive, all other things being equal. Redundancy is *not* a bad thing when developing a scale. In fact, the theoretical models that guide our scale development efforts are based on redundancy. In discussing the Spearman-Brown prophecy formula in Chapter 3, I pointed out that reliability varies as a function of the number of items, all else being equal. We are attempting to capture the phenomenon of interest by developing a set of items that reveal the phenomenon in different ways. By using multiple and seemingly redundant items, the content that is common to the items will summate across items while their irrelevant idiosyncrasies will cancel out. Without redundancy, this would be impossible.

Not all forms of redundancy are desirable, however. Useful redundancy pertains to the construct, not incidental aspects of the items. Consider two items: an original, "A really important thing is my child's success," and an altered version, "The really important thing is my child's success." Changing nothing more than an "a" to "the" in an item will certainly give you redundancy with respect to the important content of the item, but the original and altered items will also be redundant with respect to many things that you want to vary, such as their basic grammatical structure and choice of words. On the other hand, two items such as "I will do almost anything to ensure my child's success" and "No sacrifice is too great if it helps my child succeed" may be usefully redundant because they express a similar idea in somewhat different ways. They are redundant with respect to the variable of interest but not with respect to their grammatical structure and incidental vocabulary. When irrelevant redundancies are avoided, relevant redundancies will yield more reliable item sets.

Moreover, although redundancy in the final instrument may be undesirable, it is less of an issue during the early stages of item development. Consequently, even the two item versions differing by only a single word might be worth including in initial item testing. By doing so, it can be ascertained whether one or the other version is superior, and then the superior item can be incorporated into the final version of the scale. The argument against redundancy has partially been made: Redundant superficial item characteristics, such as incidental (i.e., construct-irrelevant) vocabulary or grammatical structure, are not an advantage. Construct-irrelevant similarities in wording may result in respondents' reacting similarly to items in a way that produces an inflated estimate of reliability. For example, if several items begin with a common phrase (e.g., "When I think about it, . . ."), merely sharing that phrase may cause those items to correlate more strongly with one another. A reliability indicator such as Cronbach's alpha would not distinguish between item covariation arising from that common wording and covariation attributable to the common influence of the variable of interest. Thus, reliability would be inflated.

While similar grammatical or other superficial features can constitute unwanted content similarity, redundancy that is not completely unrelated to the construct of interest can also create a problem under some circumstances. This can occur when certain items differ from most other items in a set with respect to specificity. As an example, consider a hypothetical scale intended to measure attitudes toward pet lovers. A wide variety of items might be suitable for inclusion. Other items, although relevant to the construct of interest, may be too specific—and consequently too redundant—to work well. The items "African grey parrot lovers are kind" and "I think people who like African grey parrots are good people" may be too much alike not merely because of grammatical similarity but because of potentially relevant but overly specific content that the two items share. They may pull the item set as a whole away from the intended latent variable (attitudes toward pet lovers) to an alternative, more specific latent variable (attitudes toward African grey parrot lovers). Given the wide range of pets that exist, two items about a specific and uncommon pet species are glaringly similar and likely to undermine the intent of the instrument.

More generally, how global or specific the construct of interest is can alter the impact of redundancy. Although the African grey parrot example may seem a bit extreme, inclusion of items that do not match the specificity of the construct of interest can occur in less exotic contexts. For example, in an instrument that has been designed to capture all aspects of emotion, several items about anxiety may pose a problem. Correlations among those items are likely to be greater than correlations between those items and others not about anxiety. As a consequence, these items may form a subcluster of anxiety items within the broader cluster of emotion items. This can create a number of problems. First, it may undermine the unidimensionality of the item set (which would be a problem if the investigator's intent was to develop a single measure of a unitary variable). Also, it may create an unintended focal point that results in items more similar to those in the anxiety cluster appearing to perform better than ones that are less similar. For example, although an item about *worry* and another about *fear* may be equally relevant to a broad view of emotions, the former might contribute more strongly to reliability than the latter if there is a preponderance of anxiety items in the instrument. As a result, the average correlation of the *worry* item might exceed the average correlation of the *fear* item, resulting in its stronger contribution to an estimate of reliability. De facto, an instrument including an overrepresentation of anxiety items as described would not be about all emotions equally but would be biased toward anxiety.

In contrast, the same type of anxiety items described as problematic in the preceding paragraph may not be overly redundant in an instrument with a narrower focus. Obviously, if the instrument is designed to assess anxiety, all the items should be related to that variable and that similarity would not be an instance of unwanted redundancy. In contrast, items that included a more general phrase, such as "my overall feelings," might form a subcluster if included in an anxiety scale because of their nonspecific emotional focus. What appears to be a redundancy issue, however, may actually be a matter of how well items match the specificity of the construct the investigator intends to assess.

In an instrument intended to tap a more specific variable, it is likely that the items will appear more similar to one another. Typically, for example, items in a scale measuring public speaking anxiety will appear more similar to one another (because of the specificity of the variable of interest) than items from a scale measuring emotional states more broadly. This may not be a problem as long as the similarities relate to the construct of interest. As stated earlier, items that are similar insofar as they share relevance to the *intended variable* and not in any other regard can be good items.

Number of Items

It is impossible to specify the number of items that should be included in an initial pool. Suffice it to say that you want considerably more than you plan to include in the final scale. Recall that internal consistency reliability is a function of how strongly the items correlate with one another (and hence with the latent variable) and how many items you have in the scale. As the nature of the correlations among items is usually not known at this stage of scale development, having lots of items is a form of insurance against poor internal consistency. The more items you have in your pool, the fussier you can be about choosing ones that will do the job you intend. It would not be unusual to begin with a pool of items that is three or four times as large as the final scale. Thus, a 10-item scale might evolve from a 40-item pool. If items are particularly difficult to generate for a given content area or if empirical data indicate that numerous items are not needed to attain good internal consistency, then the initial pool may be as small as 50% larger than the final scale.

In general, the larger the item pool, the better. However, it is certainly possible to develop a pool too large to administer on a single occasion to any one group of subjects. If the pool is exceptionally large, the researcher can eliminate some items based on a priori criteria, such as lack of clarity, questionable relevance, or undesirable similarity to other items.

Beginning the Process of Writing Items

Getting started writing items is often the most difficult part of the item generation process. Let me describe how I begin this process. At this point, I am less interested in item quality than in merely expressing relevant ideas. I often begin with a statement that is a paraphrase of the construct I want to measure. For example, if I were interested in developing a measure of self-perceived susceptibility to commercial messages, I might begin with "I am susceptible to commercial messages." I then would try to generate additional statements that get at the same idea somewhat differently. My next statement might be "Commercial messages affect me a lot." I would continue in this manner, imposing virtually no quality standards on the statements. My goal at this early stage is simply to identify a wide variety of ways that the central concept of the intended instrument can be stated. As I write, I may seek alternative ways of expressing critical ideas. For example, I might substitute "the things that I see in TV or magazine ads" for "commercial messages" in the next set of sentences. I find that writing quickly and uncritically is useful. After generating perhaps three or four times the number of items I anticipate including in the final instrument, I will look over what I have written. Now is the time to become critical. Items can be examined for how well they capture the central ideas and for clarity of expression. The sections that follow delineate some of the specific item characteristics to avoid or incorporate in the process of selecting from and revising the original statement list.

Characteristics of Good and Bad Items

Listing all the things that make an item good or bad is an impossible task. The content domain, obviously, has a significant bearing on item quality. However, there are some characteristics that reliably separate better items from worse ones. Most of these relate to clarity. As pointed out in Chapter 1, a good item should be unambiguous. Questions that leave the respondent in a quandary should be eliminated.

Scale developers should avoid *exceptionally lengthy items,* as length usually increases complexity and diminishes clarity. However, it is not desirable to sacrifice the meaning of an item in the interest of brevity. If a modifying clause is essential to convey the intent of an item, then include it. However, avoid unnecessary wordiness. In general, an item such as "I often have difficulty making a point" will be better than an unnecessarily longer one, such as "It is fair to say that one of the things I seem to have a problem with much of the time is getting my point across to other people."

Another related consideration in choosing or developing items is the *reading difficulty level* at which the items are written. There are a variety of methods (e.g., Dale & Chall, 1948; Fry, 1977) for assigning grade levels to passages of prose, including scale items. These typically equate longer words and sentences with higher reading levels. Reading most local newspapers presumably requires a sixth-grade reading level.

Fry (1977) delineates several steps to quantifying reading level. The first is to select a sample of text that begins with the first word of a sentence and contains exactly 100 words. (For scales having only a few items, you may have to select a convenient fraction of 100 and base subsequent steps on this proportion.) Next, count the number of complete sentences and individual syllables in the text sample. These values are used as entry points for a graph that provides grade equivalents for different combinations of sentence and syllable counts from the 100-word sample. The graph indicates that the average number of words and syllables per sentence for a fifth-grade reading level are 14 and 18, respectively. An average sentence at the sixth-grade level has 15 or 16 words and a total of 20 syllables; a seventh-grade-level sentence has about 18 words and 24 syllables. Shorter sentences with a higher proportion of longer words or longer sentences with fewer long words can yield an equivalent grade level. For example, a sentence of 9 words and 13 syllables (i.e., as many as 44% polysyllabic words) or one with 19 words and 22 syllables (i.e., no more than about 14% polysyllabic words) are both classified as sixth-grade reading level. Aiming for a reading level between the fifth and seventh grades is probably an appropriate target for most instruments that will be used with the general population. The items of the Multidimensional Health Locus of Control scales, for example, were written at a fifth- to seventh-grade reading level. A typical item at this reading level is "Most things that affect my health happen to me by accident" (Wallston et al., 1978). The item's 11 words and 15 syllables place it at the sixth-grade level.

Fry (1977) notes that semantic and syntactic factors should be considered in assessing reading difficulty. Because short words tend to be more common and short sentences tend to be syntactically simpler, his procedure is an acceptable alternative to more complex difficulty-assessment methods. However, as with other criteria for writing or choosing good items, one must use common sense in applying reading level methods. Some brief phrases containing only short words are not elementary. "Eschew casque scorn," for example, is more likely to confuse someone with a grade-school education than "Wear your helmet" will, despite the fact that both have three words and four syllables. Another source of potential confusion that should be avoided is *multiple negatives.* "I am not in favor of corporations stopping funding for anti-nuclear groups" is much more confusing than "I favor continued private support of groups advocating a nuclear ban." (It is also instructive to observe that these two statements might convey different positions on the issue. For example, the latter might imply a preference for private over public support of the groups in question.)

So-called *double-barreled* items should also be avoided. These are items that convey two or more ideas so that an endorsement of the item might refer to either or both ideas. "I support civil rights because discrimination is a crime against God" is an example of a double-barreled item. If a person supports civil rights for reasons other than its affront to a deity (e.g., because it is a crime against humanity), how should he or she answer? A negative answer might incorrectly convey a lack of support for civil rights, and a positive answer might incorrectly ascribe a motive to the respondent's support.

Another problem that scale developers should avoid is *ambiguous pronoun references.* "Murderers and rapists should not seek pardons from politicians because they are the scum of the earth" might express the sentiments of some people irrespective of pronoun reference. (However, a scale developer usually intends to be more clear about what an item means.) This sentence should be twice cursed. In addition to the ambiguous pronoun reference, it is double-barreled. *Misplaced modifiers* create ambiguities similar to ambiguous pronoun references: "Our members of Congress should work diligently to legalize prostitution in the House of Representatives" is an example of such modifiers. Using *adjective forms instead of noun forms* can also create unintended confusion. Consider the differences in meaning between "All vagrants should be given a schizophrenic assessment" and "All vagrants should be given a schizophrenia assessment."

Individual words are not the only sources of item ambiguity. Entire sentences can have more than one meaning. I have actually seen one survey of adolescent sexual behavior that included an item to assess parental education. Given the context of the survey as a whole, the wording was unfortunate: "How far did your mother go in school?" The investigators had totally failed to recognize the unintended meaning of this statement until it evoked snickers from a group of professionals during a seminar presentation. I suspect that a fair number of the adolescent respondents also got a laugh from the item. How it affected their responses to the remainder of the questionnaire is unknown.

Positively and Negatively Worded Items

Many scale developers choose to write *negatively worded items* that represent low levels or even the absence of the construct of interest as well as the more common *positively worded items,* which represent the presence of the construct of interest. The goal is to arrive at a set of items, some of which indicate a high level of the latent variable when endorsed and others that indicate a high level when not endorsed. The Rosenberg (1965) Self-Esteem Scale, for example, includes items indicative of high esteem (e.g., "I feel that I have a number of good qualities") and of low esteem (e.g., "I certainly feel useless at times"). The intent of wording items both positively and negatively within the same scale is usually to avoid an *acquiescence, affirmation,* or *agreement bias.* These interchangeable terms refer to a respondent's tendency to agree with items irrespective of their content. If, for example, a scale consists of items that express a high degree of self-esteem, then an acquiescence bias would result in a pattern of responses appearing to indicate very high esteem. If the scale is made up of equal numbers of positively and negatively worded items, on the other hand, then an acquiescence bias and an extreme degree of self-esteem could be differentiated from one another by the pattern of responses. An "agreer" would endorse items indicating both high and low self-esteem, whereas a person who truly had high esteem would strongly endorse high-esteem items and negatively endorse low-esteem items.

Unfortunately, there may be a price to pay for including positively and negatively worded items. Reversals in item polarity may be confusing to respondents, especially when completing a long questionnaire. In such a case, the respondents may become confused about the difference between expressing their strength of agreement with a statement, regardless of its polarity, versus expressing the strength of the attribute being measured (esteem, for example). As an applied social science researcher, I have seen many examples of items worded in the opposite direction performing poorly. For example, DeVellis and Callahan (1993) described a shorter, more focused alternative to the Rheumatology Attitudes Index (an unfortunate name, as the instrument does not assess attitudes and is not an index). We selected items from the original, longer version based on empirical criteria and ended up with four items expressing negative reactions to illness and one expressing the ability to cope well with illness. The intent was that users should reverse score the "coping" item so that all items expressed a sense of helplessness. More recently, Currey, Callahan, and DeVellis (2002) have examined the performance of that single item worded in the positive direction. It consistently performs poorly. When the item was reworded simply by adding the word *not* to change its valence so as to be consistent with other items, its performance improved dramatically. We suspect that, although many respondents recognized the different valence of the original item, others did not. This would result in a portion of individuals for whom the original item had positive correlations with the other four items and another portion for whom the same correlations were negative. As a consequence, for the sample as a whole, correlations of that item with the other four would be markedly diminished and, thus, would produce the type of unsatisfactory performance we observed for the original, opposite-valence item. Personal experience with community-based samples suggests to me that the disadvantages of items worded in an opposite direction outweigh any benefits.

Conclusion

An item pool should be a rich source from which a scale can emerge. It should contain a large number of items that are relevant to the content of interest. Redundancy with respect to content is an asset, not a liability. It is the foundation of internal-consistency reliability which, in turn, is the foundation of validity. Items should not involve a "package deal" that makes it impossible for respondents to endorse one part of the item without endorsing another part that may not be consistent with the first. Whether or not positively and negatively worded items are both included in the pool, their wording should follow established rules of grammar. This will help avoid some of the sources of ambiguity discussed above.

STEP 3: DETERMINE THE FORMAT FOR MEASUREMENT

Numerous formats for questions exist. The researcher should consider early on what the format will be. This step should occur simultaneously with the generation of items so that the two are compatible. For example, generating a long list of declarative statements may be a waste of time if the response format eventually chosen is a checklist composed of single-word items. Furthermore, the theoretical models presented earlier are more consistent with some response formats than with others. In general, scales made up of items that are scorable on some continuum and are summed to form a scale score are most compatible with the theoretical orientation presented in this volume. In this section, however, I will discuss common formats that depart from the pattern implied by the theoretical models discussed in Chapter 2 as well as ones that adhere to that pattern.

Thurstone Scaling

There are a number of general strategies for constructing scales that influence the format of items and response options. One method is *Thurstone scaling*. An analogy may help clarify how Thurstone scaling works. A tuning fork is designed to vibrate at a specific frequency. If you strike it, it will vibrate at that frequency and produce a specific tone. Conversely, if you place the fork near a tone source that produces the same frequency as the tuning fork, the fork will begin to vibrate. In a sense, then, a tuning fork is a "frequency detector," vibrating in the presence of sound waves of its resonant frequency and remaining motionless in the presence of all other frequencies. Imagine a series of tuning forks aligned in an array such that, as one moves from left to right along the array, the tuning forks correspond to progressively higher frequency sounds. Within the range of the tuning forks' frequency, this array can be used to identify the frequency of a tone. In other words, you could identify the tone's frequency by seeing which fork vibrated when the tone was played. A

Thurstone scale is intended to work in the same way. The scale developer attempts to generate items that are differentially responsive to specific levels of the attribute in question. When the "pitch" of a particular item matches the level of the attribute a respondent possesses, the item will signal this correspondence. Often, the signal consists of an affirmative response for items that are "tuned" to the appropriate level of the attribute and a negative response for all other items. The tuning (i.e., determination of what level of the construct each item responds to) is typically determined by having judges place a large pool of items into piles corresponding with equally spaced intervals of construct magnitude or strength.

This is quite an elegant idea. Items could be developed to correspond with different intensities of the attribute, could be spaced to represent equal intervals, and could be formatted with agree-disagree response options, for example. The investigator could give these items to respondents and then inspect their responses to see which items triggered agreement. Because the items would have been precalibrated with respect to their sensitivity to specific levels of the phenomenon, the agreements would pinpoint how much of the attribute the respondent possessed. The selection of items to represent equal intervals across items would result in highly desirable measurement properties because scores would be amenable to mathematical procedures based on interval scaling.

Part of a hypothetical Thurstone scale for measuring parents' aspirations for their children's educational and career attainments might look like the following:

1. Achieving success is the only way for my child to repay my efforts as a parent. Agree_______ Disagree_______
2. Going to a good college and getting a good job are important but not essential to my child's happiness. Agree_______ Disagree_______
3. Happiness has nothing to do with achieving educational or material goals. Agree_______ Disagree_______
4. The customarily valued trappings of success are a hindrance to true happiness. Agree_______ Disagree_______

As Nunnally (1978) points out, developing a true Thurstone scale is considerably harder than describing one. Finding items that consistently "resonate" to specific levels of the phenomenon is quite difficult. The practical problems associated with the method often outweigh its advantages unless the researcher has a compelling reason for wanting the type of calibration that it provides. Although Thurstone scaling is an interesting and sometimes suitable approach, it will not be referred to in the remainder of this text. Note, however, that methods based on item response theory, discussed in a later chapter, share many of the goals of Thurstone scales while taking a somewhat different approach to achieving them.

Guttman Scaling

A *Guttman scale* is a series of items tapping progressively higher levels of an attribute. Thus, a respondent should endorse a block of adjacent items until, at a critical point, the amount of the attribute that the items tap exceeds that possessed by the subject. None of the remaining items should be endorsed. Some purely descriptive data conform to a Guttman scale. For example, a series of interview questions might ask, "Do you smoke?" "Do you smoke more than 10 cigarettes a day?" "Do you smoke more than a pack a day?" and so on. As with this example, endorsing any specific item on a Guttman scale implies affirmation of all preceding items. A respondent's level of the attribute is indicated by the highest item yielding an affirmative response. Note that, whereas both Thurstone and Guttman scales are made up of graded items, the focus is on a single affirmative response in the former case but on the point of transition from affirmative to negative responses in the latter case. A Guttman version of the preceding parental aspiration scale might look like this:

1. Achieving success is the only way for my child to repay my efforts as a parent. Agree_______ Disagree_______
2. Going to a good college and getting a good job are very important to my child's happiness. Agree_______ Disagree_______
3. Happiness is more likely if a person has attained his or her educational and material goals. Agree_______ Disagree_______
4. The customarily valued trappings of success are not a hindrance to true happiness. Agree_______ Disagree_______

Guttman scales can work quite well for objective information or in situations where it is a logical necessity that responding positively to one level of a hierarchy implies satisfying the criteria of all lower levels of that hierarchy. Things get murkier when the phenomenon of interest is not concrete. In the case of our hypothetical parental aspiration scale, for example, the ordering may not be uniform across individuals. Whereas 20 cigarettes a day always implies more smoking than 10, responses to Items 3 and 4 in the parental aspiration scale example may not always conform to the ordering pattern of a Guttman scale. For example, a person might agree with Item 3 but disagree with Item 4. Ordinarily, agreement with Item 3 would imply agreement with Item 4, but if a respondent viewed success as a complex factor that acted simultaneously as a help and a hindrance to happiness, then an atypical pattern of responses could result.

Like Thurstone scales, Guttman scales undoubtedly have their place, but their applicability seems rather limited. With both approaches, the disadvantages and difficulties will often outweigh the advantages. It is also important to reiterate that the measurement theories discussed thus far do not always apply to these types of scales. Certainly, the assumption of equally strong causal relationships between the latent variable and each of the items would not apply to Thurstone or Guttman scale items. Nunnally and Bernstein (1994) describe briefly some of the conceptual models underlying these scales. For situations in which ordered items are particularly appropriate, models based on item response theory (discussed in Chapter 7) are potentially an appropriate choice, although implementing these methods can be quite burdensome.

Scales With Equally Weighted Items

The measurement models discussed earlier fit best with scales consisting of items that are more or less equivalent "detectors" of the phenomenon of interest—that is, they are more or less parallel (but not necessarily parallel in the strict sense of the parallel tests model). They are imperfect indicators of a common phenomenon that can be combined by simple summation into an acceptably reliable scale.

One attractive feature of scales of this type is that the individual items can have a variety of response-option formats. This allows the scale developer a good deal of latitude in constructing a measure optimally suited for a particular purpose. Some general issues related to response formatting will be examined below, as will the merits and liabilities of some representative response formats.

How Many Response Categories?

Most scale items consist of two parts: a stem and a series of response options. For example, the stem of each item may be a different declarative statement expressing an opinion, and the response options accompanying each stem might be a series of descriptors indicating the strength of agreement with the statement. For now, let us focus on the response options—specifically, the number of choices that should be available to the respondent. Some item-response formats allow the subject an infinite or very large number of options, whereas others limit the possible responses. Imagine, for example, a response scale for measuring anger that resembles a thermometer, calibrated from "no anger at all" at the base of the thermometer to "complete, uncontrollable rage" at its top. A respondent could be presented with a series of situation descriptions, each accompanied by a copy of the thermometer scale, and asked to indicate, by shading in some portion of the thermometer, how much anger the situation provoked. This method allows for virtually continuous measurement of anger. An alternative method might ask the respondent to indicate, using a number from 1 to 100, how much anger each situation provoked. This provides for numerous discrete responses. Alternatively, the format could restrict the response options to a few choices, such as "none," "a little," "a moderate amount," and "a lot," or to a simple binary selection between "angry" and "not angry."

What are the relative advantages of these alternatives? A desirable quality of a measurement scale is variability. A measure cannot covary if it does not vary. If a scale fails to discriminate differences in the underlying attribute, its correlations with other measures will be restricted and its utility will be limited. One way to increase opportunities for variability is to include lots of scale items. Another is to provide numerous response options within items. If circumstances restrict an investigator to two questions regarding anger, for example, it might be best to allow respondents more latitude in describing their level of anger. Assume that the research concerns the enforcement of nonsmoking policies in a work setting. Let us further assume that the investigators want to determine the relationship between policy and anger. If they were limited to only two questions (e.g., "How much anger do you feel when you are restricted from smoking?" and "How much anger do you feel when you are exposed to others smoking in the workplace?"), they might get more useful information from a response format that allowed subjects many gradations of response than from a binary response format (e.g., "angry" and "not angry"). For example, a 0-to-100 scale might reveal wide differences in reactions to these situations and yield good variability for the two-item scale. On the other hand, if the research team were allowed to include 50 questions about smoking and anger, simple "angry" versus "not angry" indications might yield sufficient variability when the items were added to obtain a scale score. In fact, being faced with more response options on each of 50 questions might fatigue or bore the respondents, lowering the reliability of their responses.

Another issue related to the number of response options is the *respondents' ability to discriminate meaningfully.* How fine a distinction can the typical subject make? This obviously depends on what is being measured. Few things can truly be evaluated into, say, 50 discrete categories. Presented with this many options, many respondents may use only those corresponding to multiples of 5 or 10, effectively reducing the number of options to as few as five. Differences between a response of 35 and 37 may not reflect actual difference in the phenomenon being measured. Little is gained with this sort of false precision. Although the scale's variance might increase, it may be the random (i.e., error) portion rather than the systematic portion attributable to the underlying phenomenon that is increasing. This, of course, offers no benefit.

Sometimes, the respondent's ability to discriminate meaningfully between response options will depend on the *specific wording* or *physical placement* of those options. Asking a respondent to discriminate among vague quantity descriptors, such as "several," "few," and "many," may create problems. Sometimes, the ambiguity can be reduced by the arrangement of the response options on the page. Respondents often seem to understand what is desired when they are presented with an obvious continuum. Thus, an ordering such as

Many *Some* *Few* *Very Few* *None*

may imply that "some" is more than "few" because of the ordering of these items. However, if it is possible to find a nonambiguous adjective that precludes the respondents' making assumptions based on location along a continuum, so much the better. At times, it may be preferable to have fewer response options than to have ones that are ambiguous. So, for example, it may be better in the above example to eliminate either "some" or "few" and have four options rather than five. The worst circumstance is to combine ambiguous words with ambiguous page locations. Consider the following example:

Very Helpful *Not Very Helpful*

Somewhat Helpful *Not At All Helpful*

Terms such as *somewhat* and *not very* are difficult to differentiate under the best of circumstances. However, arranging these response options as they appear above makes matters even worse. If a respondent reads down the first column and then down the second, "somewhat" appears to represent a higher value than "not very." But if a respondent reads across the first row and then across the second, the implicit ordering of these two descriptors along the continuum is reversed. Due to ambiguity in both language and spatial arrangement, individuals may assign different meanings to the two options representing moderate values, and reliability would suffer as a consequence.

Still another issue is the *investigator's ability and willingness to record a large number of values* for each item. If the thermometer method described earlier is used to quantify anger responses, is the researcher actually going to attempt a precise scoring of each response? How much precision is appropriate? Can the shaded area be measured to within a quarter of an inch? A centimeter? A millimeter? If only some crude datum—say lower, middle, or upper third—is extracted from the scale, what was the point in requesting such a precise response?

There is at least one more issue related to the number of responses. Assuming that a few discrete responses are allowed for each item, *should the number be odd or even?* Again, this depends on the type of question, the type of response option, and the investigator's purpose. If the response options are bipolar, with one extreme indicating the opposite of the other (e.g., a strong positive vs. a strong negative attitude), an odd number of responses permits equivocation (e.g., "neither agree nor disagree") or uncertainty (e.g., "not sure"); an even number usually does not. An odd number implies a central "neutral" point (e.g., neither a positive nor a negative appraisal). An even number of responses, on the other hand, forces the respondent to make at least a weak commitment in the direction of one or the other extreme (e.g., a forced choice between a mildly positive or mildly negative appraisal as the least extreme response). Neither format is necessarily superior. The researcher may want to preclude equivocation if it is felt that subjects will select a neutral response as a means of avoiding a choice. In studies of social comparison choices, for example, the investigators may want to force subjects to express a preference for information about a more advantaged or less advantaged person. Consider these two alternative formats, the first of which was chosen for a study of social comparisons among people with arthritis (DeVellis et al., 1990):

1. Would you prefer information about:

    (a) Patients who have worse arthritis than you have
    (b) Patients who have milder arthritis than you have

2. Would you prefer information about:

    (a) Patients who have worse arthritis than you have
    (b) Patients who have arthritis equally as bad as you have
    (c) Patients who have milder arthritis than you have

A neutral option such as 2b might permit unwanted equivocation. A neutral point may also be desirable. In a study assessing which of two risks (e.g., boredom vs. danger) people prefer taking, a midpoint may be crucial. The researcher might vary the chance or severity of harm across several choices between a safe, dull activity and an exciting, risky one. The point at which a respondent is most nearly equivocal about risking the more exciting activity could then be used as an index of risk taking:

Indicate your relative preference for Activity A or Activity B from the alternatives listed below by circling the appropriate phrase following the description of Activity B.
Activity A: Reading a statistics book (no chance of severe injury)

1. Activity B: Taking a flight in a small commuter plane (very slight chance of severe injury)

Strongly Prefer A *Mildly Prefer A* *No Preference* *Mildly Prefer B* *Strongly Prefer B*

2. Activity B: Taking a flight in a small open-cockpit plane (slight chance of severe injury)

Strongly Prefer A *Mildly Prefer A* *No Preference* *Mildly Prefer B* *Strongly Prefer B*

3. Activity B: Parachute jumping from a plane with a backup chute (moderate chance of severe injury)

Strongly Prefer A *Mildly Prefer A* *No Preference* *Mildly Prefer B* *Strongly Prefer B*

4. Activity B: Parachute jumping from a plane without a backup chute (substantial risk of severe injury)

Strongly Prefer A *Mildly Prefer A* *No Preference* *Mildly Prefer B* *Strongly Prefer B*

5. Activity B: Jumping from a plane without a parachute and attempting to land on a soft target (almost certain severe injury)

Strongly Prefer A *Mildly Prefer A* *No Preference* *Mildly Prefer B* *Strongly Prefer B*

The other merits or liabilities of this approach aside, it would clearly require that response options include a midpoint.

Specific Types of Response Formats

Scale items occur in a dizzying variety of forms. However, there are several ways to present items that are used widely and have proven successful in diverse applications. Some of these are discussed below.

Likert Scale

One of the most common item formats is a *Likert scale.* When a Likert scale is used, the item is presented as a declarative sentence, followed by response options that indicate varying degrees of agreement with or endorsement of the statement. (In fact, the preceding example of risk taking used a Likert response format.) Depending on the phenomenon being investigated and the goals of the investigator, either an odd or even number of response options might accompany each statement. The response options should be worded so as to have roughly equal intervals with respect to agreement. That is to say, the difference in agreement between any adjacent pair of responses should be about the same as for any other adjacent pair of response options. A common practice is to include six possible responses: "strongly disagree," "moderately disagree," "mildly disagree," "mildly agree," "moderately agree," and "strongly agree." These form a continuum from strong disagreement to strong agreement. A neutral midpoint can also be added. Common choices for a midpoint include "neither agree nor disagree" and "agree and disagree equally." There is legitimate room for discussion concerning the equivalence of these two midpoints. The first implies apathetic disinterest, while the latter suggests strong but equal attraction to both agreement and disagreement. It may very well be that most respondents do not focus much attention on subtleties of language but merely regard any reasonable response option in the center of the range as a midpoint, irrespective of its precise wording.

Likert scaling is widely used in instruments measuring opinions, beliefs, and attitudes. It is often useful for these statements to be fairly (though not extremely) strong when used in a Likert format. Presumably, the moderation of opinion is expressed in the choice of response option. For example, the statements "Physicians generally ignore what patients say," "Sometimes, physicians do not pay as much attention as they should to patients' comments," and "Once in a while, physicians might forget or miss something a patient has told them" express strong, moderate, and weak opinions, respectively, concerning physicians' inattention to patients' remarks. Which is best for a Likert scale? Ultimately, of course, the one that most accurately reflects true differences of opinion is best. In choosing how strongly to word items in an initial item pool, the investigator might profitably ask, "How are people with different amounts or strengths of the attribute in question likely to respond?" In the case of the three examples just presented, the investigator might conclude that the last question would probably elicit strong agreement from people whose opinions fell along much of the continuum from positive to negative. If this conclusion proved correct, then the third statement would not do a good job of differentiating between people with strong versus moderate negative opinions.

In general, very mild statements may elicit too much agreement when used in Likert scales. Many people will strongly agree with such a statement as "The safety and security of citizens is important." One could strongly agree with such a statement (i.e., choose an extreme response option) without holding an extreme opinion. Of course, the opposite is equally true. People holding any but the most extreme views might find themselves in disagreement with an extremely strong statement (e.g., "Hunting down and punishing wrongdoers is more important than protecting the rights of individuals"). Of the two (overly mild or overly extreme) statements, the former may be the bigger problem for two reasons. First, our inclination is often to write statements that will not offend our subjects. Avoiding offensiveness is probably a good idea; however, it may lead us to favor items that nearly everyone will find agreeable. Another reason to be wary of items that are too mild is that they may represent the absence of belief or opinion. The third of our inattentive physician items in the preceding paragraph did not indicate the presence of a favorable attitude so much as the absence of an unfavorable one. Items of this sort may be poorly suited to the research goal because we are more often interested in the presence of some phenomenon than in its absence.

In summary, a good Likert item should state the opinion, attitude, belief, or other construct under study in clear terms. It is neither necessary nor appropriate for this type of scale to span the range of weak to strong assertions of the construct. The response options provide the opportunity for gradations.

The following are examples of items in Likert response formats:

1. Exercise is an essential component of a healthy lifestyle.

1	2	3	4	5	6
Strongly Disagree	*Moderately Disagree*	*Mildly Disagree*	*Mildly Agree*	*Moderately Agree*	*Strongly Agree*

2. Combating drug abuse should be a top national priority.

1	2	3	4	5
Completely True	*Mostly True*	*Equally True and Untrue*	*Mostly Untrue*	*Completely Untrue*

Semantic Differential

The *semantic differential* scaling method is chiefly associated with the attitude research of Osgood and his colleagues (e.g., Osgood & Tannenbaum, 1955). Typically, a semantic differential is used in reference to one or more stimuli. In the case of attitudes, for example, the stimulus might be a group of people, such as automobile salesmen. Identification of the target stimulus is followed by a list of adjective pairs. Each pair represents opposite ends of a continuum, defined by adjectives (e.g., *honest* and *dishonest*). As shown in the example below, there are several lines between the adjectives that constitute the response options:

Automobile Salesmen

Honest	_____	_____	_____	_____	_____	_____	_____	Dishonest
Quiet	_____	_____	_____	_____	_____	_____	_____	Noisy

In essence, the individual lines (seven and nine are common numbers) represent points along the continuum defined by the adjectives. The respondent places a mark on one of the lines to indicate the point along the continuum that characterizes his or her evaluation of the stimulus. For example, if someone regarded auto salesmen as extremely dishonest, he or she might select the line closest to that adjective. Either extreme or moderate views can be expressed by choosing which line to mark. After rating the stimulus with regard to the first adjective pair, the person would proceed to additional adjective pairs separated by lines.

The adjectives one chooses can be either bipolar or unipolar, depending, as always, on the logic of the research questions the scale is intended to address. Bipolar adjectives each express the presence of opposite attributes, such as friendly and hostile. Unipolar adjective pairs indicate the presence and absence of a single attribute, such as friendly and not friendly.

Like the Likert scale, the semantic differential response format can be highly compatible with the theoretical models presented in the earlier chapters of this book. Sets of items can be written to tap the same underlying variable. For example, items using trustworthy/untrustworthy, fair/unfair, and truthful/untruthful as endpoints might be added to the first statement in the preceding example to constitute an "honesty" scale. Such a scale could be conceptualized as a set of items sharing a common latent variable (honesty) and conforming to the assumptions discussed in Chapter 2. Accordingly, the scores of the individual "honesty" items could be added and analyzed as described in a later section concerning the evaluation of items.

Visual Analog

Another item format that is in some ways similar to the semantic differential is the *visual analog scale.* This response format presents the respondent with a continuous line between a pair of descriptors representing opposite ends of a continuum. The individual completing the item is instructed to place a mark at a point on the line that represents his or her opinion, experience, belief, or whatever is being measured. The visual analog scale, as the term *analog* in the name implies, is a continuous scale. The fineness of differentiation in assigning scores to points on the scale is determined by the investigator. Some of the advantages and disadvantages of a continuous-response format were discussed earlier. An additional issue not raised at that time concerns possible differences in the interpretation of physical space as it relates to values on the continuum. A mark placed at a specific point along the line may not mean the same thing to different people, even when the end points of the line are identically labeled for all respondents. Consider a visual analog scale for pain such as this:

No Pain At All	______________________________	*Worst Pain I Ever Experienced*

Does a response in the middle of the scale indicate pain about half of the time, constant pain of half the possible intensity, or something else entirely? Part of the problem with measuring pain is that it can be evaluated on multiple dimensions, including frequency, intensity, and duration. Also, recollections of the worst pain a given person has ever experienced are likely to be distorted. Comparisons across individuals are further complicated by the fact that different people may have experienced different levels of "the worst pain." Of course, some of these problems reside with the phenomenon used in this example—pain (see Keefe, 2000, for an excellent discussion of pain measurement)—and not with the scale per se. However, the problem of idiosyncratic assignment of values along a visual analog scale can exist for other phenomena as well.

A major advantage of visual analog scales is that they are potentially very sensitive (Mayer, 1978). This can make them especially useful for measuring phenomena before and after some intervening event, such as an intervention or experimental manipulation, that exerts a relatively weak effect. A mild rebuke in the course of an experimental manipulation, for example, may not produce a shift on a 5-point measure of self-esteem. However, a subtle but systematic shift to lower values on a visual analog scale might occur among people in the "rebuke" condition of this hypothetical experiment. Sensitivity may be more advantageous when examining changes over time within the same individual rather than across individuals (Mayer, 1978). This may be so because, in the former case, there is no additional error due to extraneous differences between individuals.

Another potential advantage of visual analog scales when they are repeated over time is that it is difficult or impossible for subjects to encode their past responses with precision. To continue with the example from the preceding paragraph, a subject would probably have little difficulty remembering which of five numbered options to a self-esteem item he or she had previously chosen in response to a multiresponse format such as a Likert scale. Unless one of the end points of a visual analog scale were chosen, however, it would be difficult to recall precisely where a mark had been made along a featureless line. This could be advantageous if the investigator were concerned that respondents might be biased to appear consistent over time. Presumably, subjects motivated to be consistent would choose the same response after exposure to an experimental intervention as prior to such exposure. The visual analog format essentially rules out this possibility. If the post-manipulation responses departed consistently (i.e., usually in the same direction) from the premanipulation response for experimental subjects and randomly for controls, then the choice of a visual analog scale might have contributed to detecting a subtle phenomenon that other methods would have missed.

Visual analog scales have often been used as single-item measures. This has the sizable disadvantage of precluding any determination of internal consistency. With a single-item measure, reliability can be determined only by the test-retest method described in Chapter 3 or by comparison with other measures of the same attribute having established psychometric properties. The former method suffers from the problems of test-retest assessments discussed earlier, notably the impossibility of differentiating instability of the measurement process from instability of the phenomenon being measured. The latter method is actually a construct validity comparison. However, because reliability is a necessary condition for validity, one can infer the reliability if validity is in evidence. Nonetheless, a better strategy may be to develop multiple visual analog items so that internal consistency can be determined.

Numerical Response Formats and Basic Neural Processes

A study by Zorzi, Priftis, and Umilitá (2002) appearing in *Nature* suggests that certain response options may correspond to how the brain processes numerical information. According to these authors, numbers arrayed in a sequence, as with the typical Likert scale, express quantity not only in their numerical values but in their locations. They suggest that the visual line of numbers is not merely a convenient representation but corresponds to fundamental neural processes. They observed that people with various brain lesions that impair spatial perception in the visual field make systematic errors in simple, visually presented mathematical problems. The spatial anomaly and the type of errors are closely linked. Individuals who could not perceive the left visual field, when asked to indicate the midpoint between two values presented in a linear array, consistently erred "to the right." For example, when individuals were asked what would be midway between points labeled "3" and "9," errors were shifted to the right (i.e., to higher values). Reversing the scale from high to low continued to produce shifts to the right (now, lower values). When the same tasks were presented in nonvisual form (e.g., by asking what the average of 3 and 9 was), the pattern did not appear. In fact, these individuals showed no deficit in performing arithmetic when it was not presented visually. Control subjects without the visual anomaly did not show the shift pattern of those with brain lesions. The authors conclude that their work constitutes "strong evidence that the mental number line is more than simply a metaphor" and that "thinking of numbers in spatial terms (as has been reported by great mathematicians) may be more efficient because it is grounded in the actual neural representation of numbers" (p. 138). Although this study, by itself, may not warrant hard-and-fast conclusions, it provides tantalizing preliminary evidence that evaluating a linear string of numbers may correspond to fundamental neural mechanisms involved in assessing quantity. If this is truly the case, then response options presented as a row of numbers may have special merit.

Binary Options

Another common response format gives subjects a choice between *binary options* for each item. The earlier examples of Thurstone and Guttman scales used binary options ("agree" and "disagree"), although scales with equally weighted items could also have binary response options. Subjects might, for example, be asked to check off all the adjectives on a list that they think apply to themselves. Or they may be asked to answer "yes" or "no" to a list of emotional reactions they may have experienced in some specified situation. In both cases, responses reflecting items sharing a common latent variable (e.g., adjectives such as "sad," "unhappy," and "blue" representing depression) could be combined into a single score for that construct.

A major shortcoming of binary responses is that each item can have only minimal variability. Similarly, any pair of items can have only one of two levels of covariation: agreement or disagreement. Recall from Chapter 3 that the variance of a scale made up of multiple equally weighted items is exactly equal to the sum of all the elements in the covariance matrix for the individual items. With binary items, each item contributes precious little to that sum because of the limitations in possible variances and covariances. The practical consequence of this is that more items are needed to obtain the same degree of scale variance if the items are binary. However, binary items are usually extremely easy to answer. Therefore, the burden placed on the subject is low for any one item. For example, most people can quickly decide whether certain adjectives are apt descriptions of themselves. As a result, subjects often are willing to complete more binary items than ones using a format demanding concentration on finer distinctions. Thus, a binary format may allow the investigator to achieve adequate variation in scale scores by aggregating information over more items.

Item Time Frames

Another issue that pertains to the formatting of items is the specified or implied time frame. Kelly and McGrath (1988), in another volume in this series, discuss the importance of considering the temporal features of different measures. Some scales will not make reference to a time frame, implying a universal time perspective. Locus of control scales, for example, often contain items that imply an enduring belief in causality. Items such as "If I take the right actions, I can stay healthy" (Wallston et al., 1978) presume that this belief is relatively stable. This is consistent with the theoretical characterization of locus of control as a generalized rather than specific expectancy for control over outcomes (although there has been a shift toward greater specificity in later measures of locus of control beliefs—e.g., DeVellis et al., 1985). Other measures assess relatively transient phenomena. Depression, for example, can vary over time, and scales to measure it have acknowledged this point (Mayer, 1978). For example, the widely used Center for Epidemiological Studies Depression Scale (Radloff, 1977) uses a format that asks respondents to indicate how often during the past week they experienced various mood states. Some measures, such as anxiety scales (e.g., Spielberger, Gorsuch, & Lushene, 1970), are developed in different forms intended to assess relatively transient states or relatively enduring traits (Zuckerman, 1983). The investigator should choose a time frame for a scale actively rather than passively. Theory is an important guide to this process. Is the phenomenon of interest a fundamental and enduring aspect of individuals' personalities, or is it likely to be dependent on changing circumstances? Is the scale intended to detect subtle variations occurring over a brief time frame (e.g., increases in negative affect after viewing a sad movie) or changes that may evolve over a lifetime (e.g., progressive political conservatism with increasing age)?

In conclusion, the item formats, including response options and instructions, should reflect the nature of the latent variable of interest and the intended uses of the scale.

STEP 4: HAVE INITIAL ITEM POOL REVIEWED BY EXPERTS

Thus far, we have discussed the need for clearly articulating what the phenomenon of interest is, generating a pool of suitable items, and selecting a response format for those items. The next step in the process is having a group of people who are knowledgeable in the content area review the item pool. This review serves multiple purposes related to maximizing the content validity (see Chapter 4) of the scale.

First, having experts review your item pool can confirm or invalidate your definition of the phenomenon. You can ask your panel of experts (e.g., colleagues who have worked extensively with the construct in question or related phenomena) to rate *how relevant they think each item is to what you intend to measure.* This is especially useful if you are developing a measure that will consist of separate scales to measure multiple constructs. If you have been careful in developing your items, then experts should have little trouble determining which items correspond to which constructs. In essence, your thoughts about what each item measures are the hypothesis, and the responses of the experts are the confirming or disconfirming data. Even if all the items are intended to tap a single attribute or construct, expert review is useful. If experts read something into an item you did not plan to include, subjects completing a final scale might do likewise.

The mechanics of obtaining evaluations of item relevance usually involve providing the expert panel with your working definition of the construct. They are then asked to rate each item with respect to its relevance vis-à-vis the construct as you have defined it. This might entail merely rating relevance as high, moderate, or low for each item. In addition, you might invite your experts to comment on individual items as they see fit. This makes their job a bit more difficult but can yield excellent information. A few insightful comments about why certain items are ambiguous, for example, might give you a new perspective on how you have attempted to measure the construct.

Reviewers also can *evaluate the items' clarity and conciseness.* The content of an item may be relevant to the construct, but its wording may be problematic. This bears on item reliability because an ambiguous or otherwise unclear item, to a greater degree than a clear item, can reflect factors extraneous to the latent variable. In your instructions to reviewers, ask them to point out awkward or confusing items and suggest alternative wordings, if they are so inclined.

A third service that your expert reviewers can provide is *pointing out ways of tapping the phenomenon that you have failed to include.* There may be a whole approach that you have overlooked. For example, you may have included many items referring to illness in a pool of items concerned with health beliefs but failed to consider injury as another relevant departure from health. By reviewing the variety of ways you have captured the phenomenon of interest, your reviewers can help you maximize the content validity of your scale.

A final word of caution concerning expert opinion: The final decision to accept or reject the advice of your experts is your responsibility as the scale developer. Sometimes, content experts might not understand the principles of scale construction. This can lead to bad advice. A recommendation I have frequently encountered from colleagues without scale development experience is to eliminate items that concern the same thing. As discussed earlier, removing all redundancy from an item pool or a final scale would be a grave error because redundancy is an integral aspect of internal consistency. However, this comment might indicate that the wording, vocabulary, and sentence structure of the items are too similar and could be improved. Pay careful attention to all the suggestions you receive from content experts. Then make your own informed decisions about how to use their advice.

At this point in the process, the scale developer has a set of items that has been reviewed by experts and modified accordingly. It is now time to advance to the next step.

STEP 5: CONSIDER INCLUSION OF VALIDATION ITEMS

Obviously, the heart of the scale development questionnaire is the set of items from which the scale under development will emerge. However, some foresight can pay off handsomely. It might be possible and relatively convenient to include some additional items in the same questionnaire that will help in determining the validity of the final scale. There are at least two types of items to consider.

The first type of item a scale developer might choose to include in a questionnaire serves to detect flaws or problems. Respondents might not be answering the items of primary interest for the reasons you assume. There may be other motivations influencing their responses. Learning this early is advantageous. One type of motivation that can be assessed fairly easily is *social desirability.* If an individual is strongly motivated to present herself or himself in a way that society regards as positive, item responses may be distorted. Including a social desirability scale allows the investigator to assess how strongly individual items are influenced by social desirability. Items that correlate substantially with the social desirability score obtained should be considered as candidates for exclusion unless there is a sound theoretical reason that indicates otherwise. A brief and useful social desirability scale has been developed by Strahan and Gerbasi (1972). This 10-item measure can be conveniently inserted into a questionnaire.

There are other sources of items for detecting undesirable response tendencies (Anastasi, 1968). The Minnesota Multiphasic Personality Inventory (Hathaway & McKinley, 1967; Hathaway & Meehl, 1951) includes several scales aimed at detecting various response biases. In some instances, it may be appropriate to include these types of scales.

The other class of items to consider including at this stage pertain to the construct validity of the scale. As discussed in Chapter 4, if theory asserts that the phenomenon you are setting out to measure relates to other constructs, then the performance of the scale vis-à-vis measures of those other constructs can serve as evidence of its validity. Rather than mounting a separate validation effort after constituting the final scale, it may be possible to include measures of relevant constructs at this stage. The resultant pattern of relationships can provide support for claims of validity or, alternatively, provide clues if the set of items does not perform as anticipated.

STEP 6: ADMINISTER ITEMS TO A DEVELOPMENT SAMPLE

After deciding which construct-related and validity items to include in your questionnaire, you must administer them, along with the pool of new items, to some subjects. The sample of subjects should be large. How large is large? It is difficult to find a consensus on this issue. Let us examine the rationale for a large sample. Nunnally (1978) points out that the primary sampling issue in scale development involves the sampling of items from a hypothetical universe (cf. Ghiselli, Campbell, & Zedeck, 1981). In order to concentrate on the adequacy of the items, the sample should be sufficiently large to eliminate subject variance as a significant concern. Nunnally suggests that 300 people is an adequate number. However, practical experience suggests that scales have been successfully developed with smaller samples. The number of items and the number of scales to be extracted also have a bearing on the sample size issue. If only a single scale is to be extracted from a pool of about 20 items, fewer than 300 subjects might suffice.

There are several risks in using too few subjects. First, the patterns of covariation among the items may not be stable. An item that appears to increase internal consistency may turn out to be a dud when it is used on a separate sample. If items are selected for inclusion (as they very well may be) on the basis of their contribution to alpha, having a small developmental sample can paint an inaccurately rosy picture of internal consistency. When the ratio of subjects to items is relatively low and the sample size is not large, the correlations among items can be influenced by chance to a fairly substantial degree. When a scale whose items were selected under these conditions is readministered, the chance factors that made certain items look good initially are no longer operative. Consequently, the alpha obtained on occasions other than the initial development study may be lower than expected. Similarly, a potentially good item may be excluded because its correlation with other items was attenuated purely by chance.

A second potential pitfall of small sample size is that the development sample may not represent the population for which the scale is intended. Of course, this can also be the case if the development sample is large, but a small sample is even more likely to exclude certain types of individuals. Thus, a scale developer should consider both the size and composition of the development sample. A careful investigator might choose to address the generalizability of a scale across populations (or some other facet) with a G-study, as discussed in Chapter 3.

Not all types of nonrepresentativeness are identical. There are at least two different ways in which a sample may not be representative of the larger population. The first involves the level of attribute present in the sample versus the intended population. For example, a sample might represent a narrower range of the attribute than would be expected of the population. This constriction of range may also be asymmetrical so that the mean score obtained on the scale for the sample is appreciably higher or lower than one would expect for the population. Opinions regarding the appropriate legal drinking age, for example, might very well differ on a college campus from opinions on the same topic in a community at large. A mean value of the attribute that is not representative does not necessarily disqualify the sample for purposes of scale development. It may yield inaccurate expectations for scale means while still providing an accurate picture of the internal consistency the scale possesses. For example, a sample of this sort might still lead to correct conclusions about which items are most strongly interrelated.

A more troublesome type of nonrepresentativeness involves a sample that is qualitatively rather than quantitatively different from the target population. Specifically, a sample in which the relationships among items or constructs may differ from the population is reason for concern. If a sample is quite unusual, items may have a different meaning for them than for people in general. The patterns of association among items might reflect unusual attributes shared among sample members but rare in the broader community. In other words, the groupings of interrelated items that emerge (e.g., from a factor analysis) may be atypical. Stated a bit more formally, the underlying causal structure relating variables to true scores may be different if a sample is unlike the population in important ways. Consider some rather obvious examples: If the members of the chosen sample do not understand a key word that recurs among the items and has relevance to the construct, their responses may tell little or nothing about how the scale would perform under different circumstances. The word *sick* means "ill" in the United States but "nauseated" (i.e., sick to one's stomach) in England. A set of questions about illness developed for one group may have a markedly different meaning for the other. If the scale concerns a specific health problem not usually associated with nausea (e.g., arthritis), items that use the word *ill* might cluster together because of their distinct meaning if the sample were British. An American sample, on the other hand, would be unlikely to differentiate statements about being ill from other health-related items. Even within the United States, the same word can have different meanings. Among rural Southerners, for example, "bad blood" is sometimes used as a euphemism for venereal disease, whereas in other parts of the country it means animosity. If an item discussing "bad blood between relatives" performed differently among a sample of rural Southerners versus other samples, it would hardly be surprising.

The consequences of this second type of sample nonrepresentativeness can severely harm a scale development effort. The underlying structure that emerges—the patterns of covariation among items that are so important to issues of scale reliability—may be a quirk of the sample used in development. If a researcher has reason to believe that the meaning ascribed to items may be atypical among a development sample, great caution should be used in interpreting the findings obtained from that sample.

STEP 7: EVALUATE THE ITEMS

After an initial pool of items has been developed, scrutinized, and administered to an appropriately large and representative sample, it is time to evaluate the performance of the individual items so that appropriate ones can be identified to constitute the scale. This is, in many ways, the heart of the scale development process. Item evaluation is second perhaps only to item development in its importance.

Initial Examination of Items' Performance

When discussing item development, we referred to some of the qualities that are desirable in a scale item. Let us reconsider that issue. The ultimate quality we seek in an item is a high correlation with the true score of the latent variable. This follows directly from the discussion of reliability in Chapter 3. We cannot directly assess the true score (if we could, we probably would not need a scale) and, thus, cannot directly compute its correlations with items. However, we can make inferences based on the formal measurement models that have been discussed thus far. When discussing parallel tests in Chapter 2, I noted that the correlation between any two items equals the square of the correlation between either item and the true score. This squared value is the reliability of each of the items. So, we can learn about relationships to true scores from correlations among items. The higher the correlations among items are, the higher the individual item reliabilities are (i.e., the more intimately they are related to the true score). The more reliable the individual items are, the more reliable the scale that they compose will be (assuming that they share a common latent variable). So, the first quality we seek in a set of scale items is that they be *highly intercorrelated.* One way to determine how intercorrelated the items are is to inspect the correlation matrix.

Reverse Scoring

If there are items whose correlations with other items are negative, then the appropriateness of *reverse scoring* those items should be considered. Earlier, I suggested that items worded in opposite directions can pose problems. Sometimes, however, we may inadvertently end up with negatively correlated items. This might happen, for example, if we initially anticipated two separate groups of items (e.g., pertaining to happiness and sadness) but decide for some reason that they should be combined into a single group. We could then wind up with statements that relate equally to the new, combined construct (e.g., affect), but some may be positive and others negative. "I am happy" and "I am sad" both pertain to affect; however, they are opposites. If we wanted high scores on our scale to measure happiness, then we would have to ascribe a high value to endorsing the "happy" item but a low value to endorsing the "sad" item. That is to say, we would reverse score the sadness item. Sometimes, items are administered in such a way that they are already reversed. For example, subjects might be asked to circle higher numerical values to indicate agreement with a "happy" item and lower values to endorse a "sad" one. One way to do this is by having the verbal descriptors for the response options (e.g., "strongly disagree," "moderately disagree," etc.) always be in the same order for all items but having the numbers associated with them either ascend or descend, depending on the item:

1. I am sad often.

6	5	4	3	2	1
Strongly Disagree	*Moderately Disagree*	*Mildly Disagree*	*Mildly Agree*	*Moderately Agree*	*Strongly Agree*

2. Much of the time, I am happy.

1	2	3	4	5	6
Strongly Disagree	*Moderately Disagree*	*Mildly Disagree*	*Mildly Agree*	*Moderately Agree*	*Strongly Agree*

This process may confuse the subject. People may ignore the words after realizing that they are the same for all items. However, it is probably preferable to altering the order of the descriptors (e.g., from "strongly disagree" to "strongly agree" from left to right for some items and the reverse for others). Another option is to have both the verbal descriptions and their corresponding numbers the same for all items but to enter different values for certain items at the time of data coding. Changing scores for certain items at the time of coding is both tedious and potentially error prone. For every subject, every item to be reverse scored must be given the special attention involved in reverse scoring. This creates numerous opportunities for mistakes.

The easiest method for reverse scoring is to do so electronically once the data have been entered into a computer. A few computer statements can handle all the reverse scoring for all subjects' data. If the response options have numerical values and the desired transformation is to reverse the order of values, a simple formula can be used. For example, assume that a set of mood items formatted using a Likert scale was scored from 1 to 7, with higher numbers indicating agreement. Assume further that, for ease of comprehension, both positive mood and negative mood items used this same response format. However, if endorsing positive mood items is assigned a high score, then the scale is essentially a positive mood scale. Endorsing a positive mood item should result in a high value, and endorsing a negative mood item should yield a low value. This is what would be obtained if, for all negative mood items, responses of 7 were changed to 1, 6 to 2, and so forth. This type of transformation can be accomplished by creating a new score from the old score with the following formula: NEW = $(J + 1)$ – OLD, where NEW and OLD refer to the transformed and original scores, respectively, and J is the original number of response options. In the example presented, J would equal 7 and $(J + 1)$ would be 8. Subtracting a score of 7 from 8 would yield 1, subtracting 6 would yield 2, and so forth.

Some negative correlations among items may not be correctable by reverse scoring items. For example, reverse scoring a given item might eliminate some negative correlations but create others. This usually indicates that some of the items simply do not belong because they are not consistently related to other items. Any item that is positively correlated with some and negatively correlated with others in a homogeneous set should be eliminated if no pattern of reverse scoring items eliminates the negative correlations.

Item-Scale Correlations

If we want to arrive at a set of highly intercorrelated items, then each individual item should correlate substantially with the collection of remaining items. We can examine this property for each item by computing its *item-scale correlation.* There are two types of item-scale correlation. The corrected item-scale correlation correlates the item being evaluated with all the scale items, excluding itself, while the uncorrected item-scale correlation correlates the item in question with the entire set of candidate items, including itself. If there were 10 items being considered for a scale, the corrected item-scale correlation for any one of the 10 items would consist of its correlation with the other 9. The uncorrected correlation would consist of its correlation with all 10. In theory, the uncorrected value tells us how representative the item is of the whole scale. This is analogous, for example, to correlating one subset of an IQ test with the entire test to determine if the subscale is a suitable proxy. However, although an uncorrected item-total correlation makes good conceptual sense, the reality is that the item's inclusion in the scale can inflate the correlation coefficient. The fewer the number of items in the set, the bigger the difference the inclusion or exclusion of the item under scrutiny will make. In general, it is probably advisable to examine the corrected item-total correlation. An item with a high value for this correlation is more desirable than an item with a low value.

Item Variances

Another valuable attribute for a scale item is *relatively high variance.* To take an extreme case, if all individuals answer a given item identically, it will not discriminate at all among individuals with different levels of the construct being measured and its variance will be 0. In contrast, if the development sample is diverse with respect to the attribute of interest, then the range of scores obtained for an item should be diverse as well. This implies a fairly high variance. Of course, increasing variance by adding to the error component is not desirable.

Item Means

A mean *close to the center of the range* of possible scores is also desirable. If, for example, the response options for each item ranged from 1 (corresponding with "strongly disagree") to 7 (for "strongly agree"), an item mean near 4 would be ideal. If a mean were near one of the extremes of the range, then the item might fail to detect certain values of the construct. A piling up of scores at the value 7, for example, would suggest that the item was not worded strongly enough (i.e., that it was rare to find anyone who would disagree with it).

Generally, items with means too near to an extreme of the response range will have low variances, and those that vary over a narrow range will correlate poorly with other items. As stated previously, an item that does not vary cannot covary. Thus, either a lopsided mean or a low variance for any reason will tend to reduce an item's correlation with other items. As a result, you can usually concentrate primarily on the pattern of correlations among items as a gauge of their potential value. Inspecting means and variances, however, is a useful double-check once a tentative selection of items has been made on the basis of the correlations.

Factor Analysis

A set of items is not necessarily a scale. Items may have no common underlying variable (as in an index or emergent variable) or may have several. Determining the nature of latent variables underlying an item set is critical. For example, an assumption underlying alpha is that the set of items is unidimensional. The best means of determining which groups of items, if any, constitute a unidimensional set is by factor analysis. This topic is sufficiently important to merit an entire chapter (see Chapter 6). Although factor analysis requires substantial sample sizes, so does scale development in general. If there are too few respondents for factor analysis, the entire scale development process may be compromised. Consequently, factor analysis of some sort should generally be a part of the scale development process at this stage.

Coefficient Alpha

One of the most important indicators of a scale's quality is the reliability coefficient, alpha. Virtually all the individual-item problems discussed thus far—a noncentral mean, poor variability, negative correlations among items, low item-scale correlations, and weak inter-item correlations—will tend to reduce alpha. Therefore, after we have selected our items—weeding out the poor ones and retaining the good ones—alpha is one way of evaluating how successful we have been. Alpha is an indication of the proportion of variance in the scale scores that is attributable to the true score. There are several options for computing alpha, differing in degree of automation. Some computer packages have item analysis programs that compute alpha. In SPSS, the RELIABILITY procedure computes alpha for a full scale and for all $k - 1$ versions (i.e., for every possible version with a single item removed). The program also provides corrected and uncorrected item-scale correlations. SAS includes alpha calculations as a feature of the correlation program, PROC CORR. By including the option ALPHA in the PROC CORR statement, the variables listed in the VAR statement will be treated as a scale and alpha will be computed for the full set of items as well as all possible $k - 1$ item sets. Item-scale correlations are also provided.

Another option for computing alpha is to do so by hand. If variances for the individual items and for the scale as a whole are available, they can be plugged into the first formula for alpha discussed in Chapter 3. Or one can use the Spearman-Brown formula, which was also introduced in Chapter 3. This formula uses information available from a correlation matrix rather than variances as the basis for computing alpha. A shortcoming of this approach is that correlations are standardized covariances, and standardizing the individual items might affect the value of alpha. If one adheres strictly to the model of parallel tests, then this is inconsequential because the correlations are assumed to be equal. However, they virtually never are exactly equal. The essentially tau-equivalent tests model does not require equal correlations among items, only equal covariances. Thus, the proportion of each individual item's variance that is due to error is free to vary under that model. However, because the Spearman-Brown formula actually works with *average* inter-item correlations, and one of the implications of the tau-equivalent model is that the average item-scale correlations are equal for each item, there is still no problem. Nonetheless, there can be small (but sometimes large) differences between the values of alpha obtained from covariance-based versus correlation-based computational methods. Because the covariance matrix uses the data in a purer form (without standardization), it is preferred and should generally be used.

Theoretically, alpha can take on values from 0.0 to 1.0, although it is unlikely that it will attain either of these extreme values. If alpha is negative, something is wrong. A likely problem is negative correlations (or covariances) among the items. If this occurs, try reverse scoring or deleting items as described earlier in this chapter. Nunnally (1978) suggests a value of .70 as an acceptable lower bound for alpha. It is not unusual to see published scales with lower alphas. Different methodologists and investigators begin to squirm at different levels of alpha. My personal comfort ranges for research scales are as follows: below .60, unacceptable; between .60 and .65, undesirable; between .65 and .70, minimally acceptable; between .70 and .80, respectable; between .80 and .90, very good; and much above .90, one should consider shortening the scale (see the following section). I should emphasize that these are *personal and subjective* groupings of alpha values. I cannot defend them on strictly rational grounds. However, they reflect my experience and seem to overlap substantially with other investigators' appraisals. The values I have suggested apply to *stable* alphas. During development, items are selected, either directly or indirectly, on the basis of their contribution to alpha. Some of the apparent covariation among items may be due to chance. Therefore, it is advisable during the development stage to strive for alphas that are a bit higher than you would like. Then, if the alphas deteriorate somewhat when used in a new research context, they will still be acceptably high. As noted earlier, if the developmental sample is small, the investigator should be especially concerned that the initial alpha estimates obtained during scale development may not be stable. As we shall see, this is also the case when the number of items making up the scale is small.

A situation in which the suggested "comfort ranges" for alpha do not apply is when one is developing a scale that requires critical accuracy. Clinical situations are an example. The suggested guidelines are suitable for *research instruments* that will be used with *group data.* For example, a scale with an alpha of .85 is probably perfectly adequate for use in a study comparing groups with respect to the construct being measured. Individual assessment, especially when important decisions rest on that assessment, demand a much higher standard. Scales that are intended for individual diagnostic, employment, academic placement, or other important purposes should probably have considerably higher reliabilities, in the mid-.90s, for example.

In some situations, such as when a scale consists of a single item, it will be impossible to use alpha as the index of reliability. If possible, some reliability assessment should be made. Test-retest correlation may be the only option in the single-item instance. Although this index of reliability is imperfect, as discussed in Chapter 3, it is clearly better than no reliability assessment at all. A preferable alternative, if possible, would be to constitute the scale using more than a single item.

STEP 8: OPTIMIZE SCALE LENGTH

Effect of Scale Length on Reliability

At this stage of the scale development process, the investigator has a pool of items that demonstrate acceptable reliability. A scale's alpha is influenced by two characteristics: the extent of covariation among the items and the number of items in the scale. For items that have item-scale correlations about equal to the *average* inter-item correlation (i.e., items that are fairly typical), adding more will increase alpha and removing more will lower it. Generally, shorter scales are good because they place less of a burden on respondents. Longer scales, on the other hand, are good because they tend to be more reliable. Obviously, maximizing one of these assets reduces the other. Therefore, the scale developer should give some thought to the optimal trade-off between brevity and reliability.

If a scale's reliability is too low, then brevity is no virtue. Subjects may, indeed, be more willing to answer a 3-item scale than a 10-item scale. However, if the researcher cannot assign any meaning to the scores obtained from the shorter version, then nothing has been gained. Thus, the issue of trading reliability for brevity should be confined to situations when the researcher has "reliability to spare." When this is, in fact, the case, it may be appropriate to buy a shorter scale at the price of a little less reliability.

Effects of Dropping "Bad" Items

Whether dropping "bad" items actually increases or slightly lowers alpha depends on just how poor the items are that will be dropped and on the number of items in the scale. Consider the effect of more or fewer items that are equally "good" (i.e., that have comparable correlations with their counterparts): With fewer items, a greater change in alpha results from the addition or subtraction of each item. If the average inter-item correlation among four items is .50, the alpha will equal .80. If there are only three items with an average inter-item correlation of .50, alpha drops to .75. Five items with the same average correlation would have an alpha of .83. For 9-, 10-, and 11-item scales with average inter-item correlations of .50, alphas would be .90, .91, and .92, respectively. In the latter instances, the alphas are not only higher but much closer in value to one another.

If an item has a sufficiently lower-than-average correlation with the other items, dropping it will raise alpha. If its average correlation with the other items is only slightly below (or equal to or above) the overall average, then retaining the item will increase alpha. I stated above that a four-item scale would attain an alpha of .80, with an average inter-item correlation of .50. How low would the average correlation of one item to the other three have to be for that item's elimination to help rather than hurt alpha? First, consider what the average inter-item correlation would have to be for a three-item scale to achieve an alpha of .80. It would need to be .57. So, after eliminating the worst of four items, the remaining three would need an average inter-item correlation of .57 for alpha to hold its value of .80. Three items whose average inter-item correlation was lower than .57 would have a lower alpha than four items whose inter-item correlations averaged .50. Assuming that the three best items of a four-item scale had an average correlation among them of .57, the average correlation between the remaining (and, thus, worst) item and the other three would have to be lower than .43 for its elimination to actually increase alpha. (Having three items whose intercorrelations average .57 and one whose average correlation with the other three is .43 yields an overall average inter-item correlation among the four of .50.) For any value larger than .43, having a fourth item does more good than lowering the average inter-item correlation does harm. Thus, the one "bad" item would have to be a fair bit worse than the other three (.57 – .43 = .14) to be worth eliminating.

Now, consider the situation when there is a 10-item scale with an alpha of .80. First of all, the average inter-item correlation need only be about .29, illustrating the manner in which more items offset weaker correlations among them. For a nine-item scale to achieve the same alpha, the average inter-item correlation would need to be about .31. A "bad" item would need to have an average inter-item correlation with the remaining nine items of about .20 or less in order for its inclusion as a tenth item to pull the overall average inter-item correlation below .29. A failure to bring the average below this value would result in the item's inclusion benefitting alpha. The average inter-item correlation difference between the nine "good" items and the one "bad" in this case is .31 – .20 = .11, a smaller difference than the one found in the four-item example.

Tinkering With Scale Length

How does one go about "tinkering" with scale length in practice? Obviously, items that contribute least to the overall internal consistency should be the first to be considered for exclusion. These can be identified in a number of ways. The SPSS RELIABILITY procedure and the ALPHA option of PROC CORR in SAS show what the effect of omitting each item would be on the overall alpha. The item whose omission has the least negative or most positive effect on alpha is usually the best one to drop first. The item-scale correlations can also be used as a barometer of which items are expendable. Those with the lowest item-scale correlations should be eliminated first. SPSS also provides a squared multiple correlation for each item, obtained by regressing the item on all the remaining items. This is an estimate of the item's *communality,* the extent to which it shares variance with the other items. As with item-scale correlations, items with the lowest squared multiple correlations should be the prime candidates for exclusion. Generally, these various indices of item quality converge. A poor item-scale correlation is typically accompanied by a low squared multiple correlation and a small loss, or even a gain, in alpha when the item is eliminated. Scale length affects the precision of alpha. In practice, a computed alpha is an estimate of reliability dependent on the appropriateness of the measurement assumptions to the actual data. It has already been noted that alpha increases when more items are included (unless they are relatively poor items). In addition, the *reliability of alpha as an estimate of reliability* increases with the number of items. This means that an alpha computed for a longer scale will have a narrower confidence interval around it than will an alpha computed for a shorter scale. Across administrations, a longer scale will yield more similar values for alpha than will a shorter one. This fact should be considered in deciding how long or short to make a scale during development.

Finally, it is important to remember that a margin of safety should be built into alpha when trying to optimize scale length. Alpha may decrease somewhat when the scale is administered to a sample other than the one used for its development.

Split Samples

If the development sample is sufficiently large, it may be possible to split it into two subsamples. One can serve as the primary development sample, and the other can be used to cross-check the findings. So, for example, data from the first subsample can be used to compute alphas, evaluate items, tinker with scale length, and arrive at a final version of the scale that seems optimal. The second subsample can then be used to replicate these findings. The choice of items to retain will not have been based at all on the second subsample. Thus, alphas and other statistics computed for this group would not manifest the chance effects, such as alpha inflation, that were discussed earlier. If the alphas remain fairly constant across the two subsamples, you can be more comfortable assuming that these values are not distorted by chance. Of course, the two subsamples are likely to be much more similar than two totally different samples. The subsamples, divided randomly from the entire development sample, are likely to represent the same population; in contrast, an entirely new sample might represent a slightly different population. Also, data collection periods for the two subsamples are not separated by time, whereas a development sample and a totally separate sample almost always are. Furthermore, any special conditions that may have applied to data collection for one subsample would apply equally to the other. Examples of special conditions include exposure to specific research personnel, physical settings, and clarity of questionnaire printing. Also, the two subsamples may be the only two groups to complete the scale items together with all the items from the original pool that were eventually rejected. If rejected items exercised any effects on the responses to the scale items, these would be comparable for both subsamples.

Despite the unique similarity of the resultant subsamples, replicating findings by splitting the developmental sample provides valuable information about scale stability. The two subsamples differ in one key aspect: In the case of the first subsample on whose data item selection was based, the opportunity existed for unstable, chance factors to be confused with reliable covariation among items. No such opportunity for systematically attributing chance results to reliability exists for the second group because its data did not influence item selection. This crucial difference is sufficient reason to value the information that sample splitting at this stage of scale development can offer.

The most obvious way to split a sufficiently large sample is to halve it. However, if the sample is too small to yield adequately large halves, you can split unevenly. The larger subsample can be used for the more crucial process of item evaluation and scale construction and the smaller for cross-validation.

EXERCISES

Assume that you are developing a fear-of-snakes measure that has a six-choice Likert response format and uses 300 subjects. Although more items would be desirable for actual scale development, complete the following steps for these exercises:

1. Generate a pool of 10 Likert-format items.
2. For each item you have written, estimate what Likert scale values would be endorsed by the "average person" (i.e., neither a snake phobic nor a snake charmer).
3. Pick an item from the pool that you suspect might elicit an extreme response from an average person and rewrite it to elicit a more moderate response.
4. Generate another 10 Likert items to tap a construct *other than* fear of snakes. Randomly mix these items with the original 10 and ask a few friends to indicate what they think each of the items is intended to measure.
5. Using either fear of snakes or the construct underlying your second pool of 10 items, list directly observable behaviors that could be used to validate a scale measuring that construct and explain how you could use behavioral data for validation.
6. What would the alpha for the scale be if your 10 fear-of-snakes items had an average inter-item correlation of .30?[1]
7. How could you use split samples to estimate and cross-validate the scale's coefficient alpha?

NOTE

1. Answer: Alpha = $[10 \times .30] / [1 + (9 \times .30)] = .81$.

6

Factor Analysis

In Chapter 2, when discussing different theoretical models that could describe the relationship of a scale's items to the latent variable, I mentioned the general factor model. That model does not assume that only one latent variable is the source of all covariation among the items. Instead, that model allows multiple latent variables to serve as causes of variation in a set of items.

To illustrate how more than one latent variable might underlie a set of items, I will revisit a specific, albeit hypothetical, situation that I have mentioned in earlier chapters. Many constructs of interest to social and behavioral scientists can be operationalized at multiple levels of specificity. The terms *psychological adjustment, affect, negative affect, anxiety, social anxiety,* and *public speaking anxiety* are examples of hierarchical phenomena. Each term could subsume those that follow it in the list, and it might be possible to develop measures at each level of specificity. Presumably, differently worded items with different time frames and response options could tap either a specific, middling, or general level of this continuum. Hopefully, a scale developer would select item wordings that corresponded with the intended level of variable specificity. Factor analysis then could be used to assess whether that selection process succeeded.

To make this example more specific, consider a set of 25 items that all pertain to affect. Recall that one of the assumptions of classical measurement theory is that items composing a scale are unidimensional—that is, they tap only a single construct. Our concern, therefore, is whether these 25 items should make up one general or many more specific scales. Do all 25 items belong together? Or is it more appropriate to have separate scales for different affective states, such as depression, euphoria, hostility, anxiety, and so on? Maybe it would be even better to split the positive and negative affect items (e.g., "happy" vs. "sad" for depression or "tense" vs. "calm" for anxiety) into separate scales. How do we know what is most appropriate for the items at hand? Essentially, the question is, does a set of items asking about several affective states have one or many latent variables underlying it?

Attempting to answer these questions while excluding factor analysis and relying only on the methods discussed in the preceding chapters would be daunting. We could compute alpha on the entire set of mood items. Alpha would tell us something about how much variance a group of items had in common. If alpha were low, we might search for subsets of items correlating more strongly with each other. For example, we might suspect that positive and negative affect items do not correlate with each other and that combining them was lowering alpha. The alphas for these more homogeneous (all positive or all negative affect) subsets of items should be higher. We might then speculate that even more homogeneous subsets (e.g., separating anxiety from depression in addition to positive from negative) should have still higher alphas. However, at some point, we might also worry that these more specific and homogeneous scales would correlate strongly with each other because they were merely tapping different aspects of the same affective state. This would suggest that their items belonged in the same rather than in separate scales.

It should be emphasized that a relatively high alpha is no guarantee that all the items reflect the influence of a single latent variable. If a scale consisted of 25 items, 12 reflecting primarily one latent variable and the remaining 13 primarily another, the correlation matrix for all the items should have some high and some low values. Correlations based on two items representing primarily the same latent variable should be high, and those based on items primarily influenced by different latent variables should be relatively low. However, the *average* inter-item correlation might be high enough to yield a respectable alpha for a 25-item scale. In fact, to yield an alpha of .80 for a scale of that length, the average inter-item correlation needs only to be .14. Thus, with items measuring the same construct correlating about .29 (a respectable but certainly not unusually large average correlation for items within a homogeneous subset) and those between sets correlating about .00, one could still achieve an alpha of about .80. But, of course, alpha is appropriate only for a unidimensional item set, so this apparently adequate reliability score would be misleading as an indicator of either internal consistency or unidimensionality (which we know does not apply in this hypothetical example).

Factor analysis, the topic of this chapter, is a useful analytic tool that can tell us, in a way that reliability coefficients cannot, about important properties of a scale. It can help us determine *empirically* how many constructs, or latent variables, or *factors* underlie a set of items.

OVERVIEW OF FACTOR ANALYSIS

Factor analysis serves several related purposes. One of its primary functions, as just noted, is to help an investigator in *determining how many latent variables* underlie a set of items. Thus, in the case of the 25 affect items, factor analysis could help the investigator determine whether one broad or several more specific constructs were needed to characterize the item set. Factor analysis also can provide a means of explaining variation among relatively many original variables (e.g., 25 items) using relatively few newly created variables (i.e., the factors). This amounts to *condensing information* so that variation can be accounted for by using a smaller number of variables. For example, instead of needing 25 scores to describe how respondents answered the items, it might be possible to compute fewer scores (perhaps even one) based on combining items. A third purpose of factor analysis is *defining the substantive content or meaning of the factors* (i.e., latent variables) that account for the variation among a larger set of items. This is accomplished by identifying groups of items that covary with one another and appear to define meaningful underlying latent variables. If, say, two factors emerged from an analysis of the 25 affect items, the content of individual items making up those factor groupings could provide a clue about the underlying latent variables represented by the factors. A fourth function of factor analysis is related to all three of the previously mentioned functions. Factor analysis aids in *identifying items that are performing better or worse.* Thus, individual items that do not fit into any of the factorially derived categories of items or fit into more than one of those categories can be identified and considered for elimination.

The following sections present a conceptual summary of factor analysis. Readers who want a more thorough computational treatment of factor analysis should consult a text devoted to the topic, such as Cureton (1983), Gorsuch (1983), Harman (1976), or McDonald (1984).

Examples of Methods Analogous to Factor Analytic Concepts

To get an intuitive sense of what factor analysis does, we can consider two less formal but roughly analogous processes with which we may be more familiar. The first of these processes is sometimes used in human resources management to identify common themes among seemingly diverse specific issues that may concern team members or coworkers.

Example 1

Assume that a small, new company wants to identify what characteristics its employees believe are important for their coworkers to have. They believe that identifying and rewarding widely valued characteristics will play an important part in cultivating a harmonious and cooperative work environment. The company hires a human resources specialist to assist them. This person, whom we will call Jim, gathers the company's 10 employees together and explains that he would like them to consider what characteristics of their fellow employees they regard as important across the range of interactions they might have on the job, from developing proposals and reports together to interacting with potential clients together to sharing a table in the cafeteria—the full range of interactions employees might have. Jim suggests that, to begin the process, the employees each write on separate pieces of paper as many important characteristics as they can identify.

After several minutes, during which employees write down their ideas, Jim asks for a volunteer to read one of his or her ideas to the group. Alice says that one characteristic she wrote down was "willing to share ideas." Jim thanks her and asks that she tape the paper with that idea written on it to the wall. Another employee, Bill, reads one of his characteristics: "a sense of humor." This, too, is taped to the wall. The process continues with each individual employee stating each of the characteristics he or she had written down. In this way, people individually name a variety of characteristics that they personally consider important in coworkers. After doing so, they tape the sheets of paper naming each characteristic to the wall. Among the characteristics listed are the following:

willing to share ideas
sense of humor
always has the right tools for the job
smart
isn't sloppy
hard worker
thinks logically
comes through in a pinch
prepares for tasks
makes good impression with clients
doesn't try to get all the credit
team player
fun
has a nice car
has a lot of experience in this type of work
is friendly
can be counted on
pays attention to details
has a mind like a steel trap
is outgoing
knows a lot of potential clients
is reliable
has character
is well educated
is trustworthy
knows how to dress
is a good storyteller
is intelligent
is a person of faith
is willing to work long hours if that's what it takes to get the job done

This process continues for some time, and soon the wall is covered with more than 30 slips of paper, each naming a characteristic that an employee thought was important. Next, Jim asks if people see any characteristics that they think go together. Katherine points out that "smart" and "is intelligent" are the same. Jim takes the sheet of paper on which "is intelligent" is written and moves it next to "smart." Frank suggests that "is well educated" should also be part of that group. Several other characteristics are added to the same group of statements. Then Carla observes that "is friendly" and "makes a good impression with clients" are similar to each other and different from the other group of statements already formed. She suggests that those two characteristics should be put together in a new group. Then, "fun" is also added to this second group. "Isn't sloppy" and "knows how to dress" form the kernel of a third group until one employee says she thinks "isn't sloppy" would go better with "prepares for tasks" than with "knows how to dress." This process continues until Jim and the employees have formed several clusters of statements. Virtually every characteristic described gets placed into some group.

Jim then asks people to give names—a word or short descriptive phrase—to each group of statements. Various groups of items are labeled "Intelligence," "Appearance," "Conscientiousness," "Personality," "Dependability," and other category names. Presumably, each group of statements represents a key concept related to employees' perceptions of one another's characteristics.

Example 2

Several years later, the company decides to repeat the exercise. The managers suspect that things have changed enough that not all the categories originally identified may still be relevant. Jim, the human resources facilitator, is unavailable. Carol, one of the company's executives, decides that an easier way of getting at similar information might be to develop a questionnaire that has statements like the ones people came up with in the earlier exercise. Employees would be asked to indicate how important they felt each characteristic was, using "not at all," "somewhat," and "very" as their response options. These questionnaires were administered to the employees, who now numbered nearly 150. When Carol got the questionnaires back, she looked them over to see which things were most important. One thing she noticed was that different things were important to different people but certain characteristics tended to be rated similarly to one another. For example, people who thought "pays attention to details" was important were likely to consider "prepares for tasks" important as well. People who did not consider one of these items important typically did not consider the other important, either. Carol mulled over the pile of questionnaires and thought about how to make sense of them. She remembered how, during the original exercise conducted several years earlier with the slips of paper on the wall, there seemed to be more groups than were really needed. Some of the individual statements were fairly worthless, she thought, and sometimes if the same person had more than one of these worthless statements, a whole worthless category would result. She wondered if there was a way of determining how many categories it would take to distill most of what the employees thought about their coworkers. As an exercise, she tried looking for other sets of items, like the two she had noticed already, that tended to be endorsed similarly across employees. In essence, she looked for groupings of similar items based not on just their content but on how similarly they were evaluated by employees. This took a great deal of time, and Carol was not really sure if she was picking up on all the important clusters of statements, but she felt able to garner some interesting ideas in this way from the questionnaires.

Shortcomings of These Methods

Both of these examples are conceptually analogous to factor analysis but with certain important differences. In both cases, the result is a reorganization of a substantial amount of specific information into a more manageable set of more general but meaningful categories. Presumably, each of these reclassifications resulted in a few ideas that captured much of what the many individual statements covered. Both approaches had fairly obvious shortcomings, however.

In the first example, there was little control over the quality of the statements generated. This shortcoming arises from the fact that the people developing the items will vary in how skilled they are at capturing relevant ideas in concise phrases. Some people are more extroverted than others and may end up generating more statements. It is not always the case, however, that the most extroverted group members will be the most insightful. For this and other reasons, this process often leads to statements that are ambiguous, irrelevant, or downright silly. Removing or improving poor items may be difficult. Depending on the dynamics of the group involved in the exercise, it may be awkward to dismiss such statements without offending their authors. Consequently, they end up being given as much credibility as better statements reflecting more germane ideas. Despite this unevenness of quality from item to item, all may tend to be treated more or less equally.

If several silly but similar items are offered, they are likely to constitute a category based merely on their similarity. Categories may be prioritized, but this is usually done by consensus and, depending on who generated what statements, there may be reluctance to brand certain categories as trivial. My experience with exercises of this sort suggests, furthermore, that there is a strong tendency to place every statement in *some* category. Having several neat categories and then one or two orphaned statements seems to leave people with a lack of closure so that orphan statements find a home, even if the fit is not immediately evident. Furthermore, even when one can identify which specific statements exemplify a category, it is not necessarily obvious which are better or worse exemplars.

So how did this method fare overall with respect to the functions of factor analysis identified earlier? Clearly, this method made it difficult to *identify items that are performing better or worse.* Item quality control is at odds with the collective nature of this task and might risk offending some participants. Although the method just described has some utility in *determining how many latent variables* underlie the statements that employees generated, basing that determination on the subjective impressions of the employees, as it does, seems less than ideal. Although the employees may have valuable insights into their workplace, it is less likely that they have experience in organizing those insights into coherent groups representing important underlying constructs. Moreover, there is little objective basis for determining whether the employees as a group have done a good or bad job of determining the number of latent variables underlying the statements generated, because the process relies solely on subjective criteria. With regard to *condensing information,* a similar problem arises. Although some items (or even whole categories) may be discarded, the criteria used are necessarily subjective and there is little means for judging objectively whether good choices have been made. The one respect in which this method may not do too badly is *defining the substantive content or meaning of the factors.* If the factors are credible (which is open to question), examining the content of statements offered may give insights into the underlying constructs that employees had in mind as important.

The second example avoided some of these shortcomings. Carol could weed out items that struck her as irrelevant, although this placed quite a burden on her judgment. At least the process of endorsing items was somewhat more democratic. Every person got to evaluate every item without risk of alienating a coworker. Thus, the process of *identifying items that are performing better or worse* may at least have been more consistent. Furthermore, the groupings were determined not by a mere sense of apparent statement similarity but by evidence that people reacted to similarly grouped items in a common way. That is, the similarity was a characteristic of the items (certain sets of which seemed to elicit similar perceptions), not the respondents (who varied in their responses to any specific item). Seeing one item in a group as unimportant implies a substantial likelihood of the same individual seeing the other items in the same group as unimportant. A different employee might see those same items as consistently important. The critical issue is that, whatever an individual's assessment of importance, it tended to be consistent across statements within a group. In fact, that was the basis on which Carol constituted the groups. Thus, the process of *determining how many latent variables* there were may have been improved somewhat by adopting this revised, questionnaire-based methodology. Of course, evaluating items by visual inspection for 150 questionnaires would be fairly daunting, and it is likely that Carol's categorization system was not the most efficient possible method. Relevant questions arise: How much consistency was required for items to be considered a group? How many instances of a single employee giving divergent assessments (i.e., an agreement and a disagreement of importance) for two items in the same potential cluster would Carol tolerate? How well this method achieved the remaining functions—*condensing information* and *defining the substantive content or meaning of the factors*—is hard to say. Both would depend on how good a job Carol was able to do at finding and dropping poor items, correctly determining what the important themes were, and then using the items within clusters to interpret what the underlying constructs might be.

In summary, these two methods clearly leave a lot to be desired if the goal is to achieve the functions mentioned above that are commonly associated with factor analysis. In fairness, the goal of the methodologies described (especially the first) may not have been to achieve those functions, but some objective roughly analogous to what factor analysis accomplishes does seem to motivate exercises like the first one described. While it is admittedly a straw man, the employee exercise as a comparison to factor analysis does serve the purpose of (a) providing a more intuitive illustration of what factor analysis can accomplish and (b) highlighting the potential shortcomings of an informal, subjective approach to identifying underlying variables.

CONCEPTUAL DESCRIPTION OF FACTOR ANALYSIS

Factor analysis is a category of procedures that accomplish the same type of classification as the methods described above but do so in accordance with a more structured set of operations and provide more explicit information that the data analyst can use to make judgments. Like the methods just described, factor analysis identifies categories of similar statements. The factor analyst's first task is to determine how many categories are sufficient to capture the bulk of the information contained in the original set of statements.

Extracting Factors

One can view factor analysis as beginning from the premise that one big category containing all the items is all that is needed (i.e., that one concept or category is sufficient to account for the pattern of responses). It then assesses how much of the association among individual items that single concept can explain. The analysis then performs a check to see how well the single-concept premise has fared. If it appears that one concept or category has not done an adequate job of accounting for covariation among the items, the factor analysis rejects the initial premise. It then identifies a second concept (i.e., latent variable or factor) that explains some of the remaining covariation among items. This continues until the amount of covariation that the set of factors has not accounted for is acceptably small.

The First Factor

How is this accomplished? The process begins with a correlation matrix for all the individual items. Using this matrix as a starting point, factor analysis examines the patterns of covariation represented by the correlations among items. What follows is a conceptual description. Certain mathematical details are omitted in the interest of clarity, so this should not be taken literally as the set of operations underlying computer-generated factor analyses.

As stated earlier, the process involves the initial premise of a single concept that can adequately account for the pattern of correlations among the items. This amounts to a provisional assertion that a model having a single latent variable (i.e., a single factor), with a separate path emanating from it to each of the items, is an accurate representation of causal relationships (see Figure 6.1). This further implies that such a model can account for the correlations among the items. To test this assumption conceptually, the factor analysis program must determine the correlation of each item with the factor representing the single latent variable and then see if the observed correlations between items can be re-created by appropriately multiplying the paths linking each pair of variables via the factor. But how can the program compute correlations between observed item responses and a factor representing a latent variable that has not been directly observed or measured?

One approach is to posit that the sum of all the item responses is a reasonable numerical estimate of the one, all-encompassing latent variable that is assumed to account for inter-item correlations. In essence, this overall sum is an estimate of the latent variable's "score." Because the actual scores for all items are presumed for the time being to be determined by one latent variable, a quantity combining information from all items (i.e., an overall sum) is a reasonable estimate of that latent variable's numerical value. It is fairly simple to add the individual item scores together into a total score and to compute item-total correlations for each separate item with the total of all items. These item-total correlations serve as proxies for the correlations between the observed items and the unobserved latent variable (i.e., the causal pathways from latent variable to individual items). With values thus assigned to those

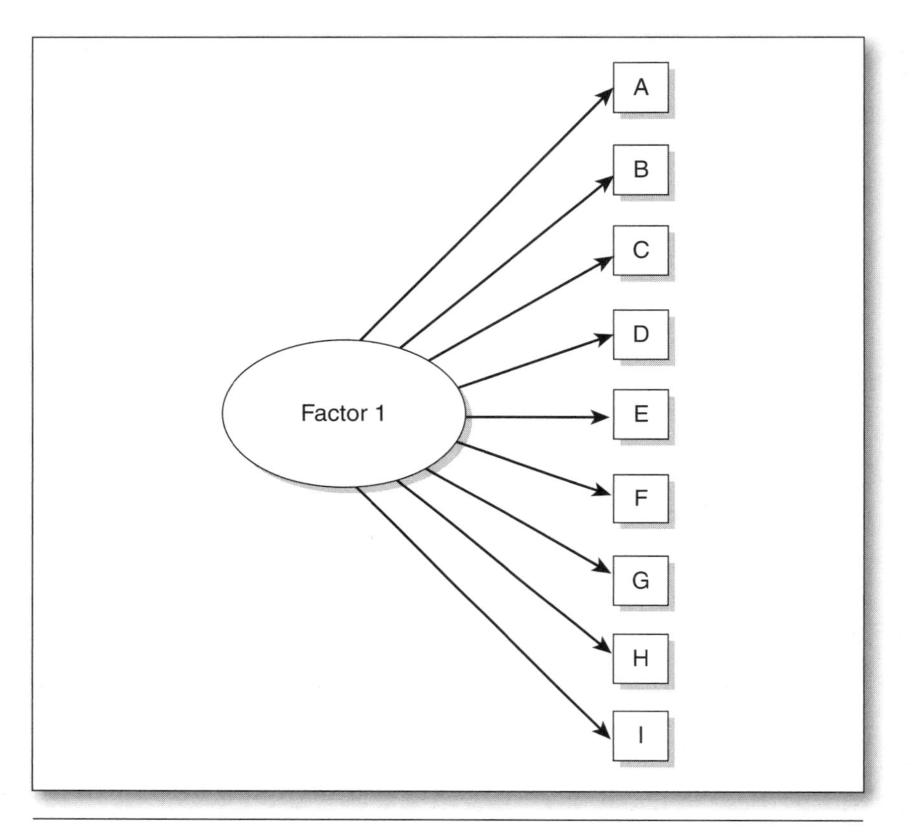

Figure 6.1 A single-factor model

causal pathways, one then can compute projected inter-item correlations based on this one-factor model. These model-derived correlations are projections of what the actual inter-item correlations should be if the premise of only one underlying variable is correct. The legitimacy of the premise can be assessed by comparing the projected correlations with the actual correlations. This amounts to subtracting each projected correlation from the corresponding actual correlation based on the original data. A substantial discrepancy between actual and projected correlations indicates that the single-factor model is not adequate; there is still some unaccounted-for covariation among the items.

It may help to consider this sequence for a single pair of items, A and B, that are part of a larger set. First, the whole set of items, including A and B, would be added together to get a summary score. Then, correlations of A with that total score and B with that total score would be computed. These two item-total correlations are assumed to represent the correlations of A and of B with the factor, which corresponds with the underlying latent variable. If the premise of a single underlying latent variable were correct, then a path diagram involving A, B, and the factor would have paths (a and b in Figure 6.2) from the latter to each of the former. The values of these paths would be the item-total correlations just described. Based on this path diagram, the correlation between A and B should be the product of those two paths. Computing this proposed correlation between A and B entails simple multiplication. Once computed, the proposed correlation can be compared with the actual correlation between A and B. The proposed correlation can be subtracted from the actual correlation to yield a residual correlation. A substantial residual correlation would indicate that invoking a single underlying latent variable as the sole cause of covariation between A and B is not adequate.

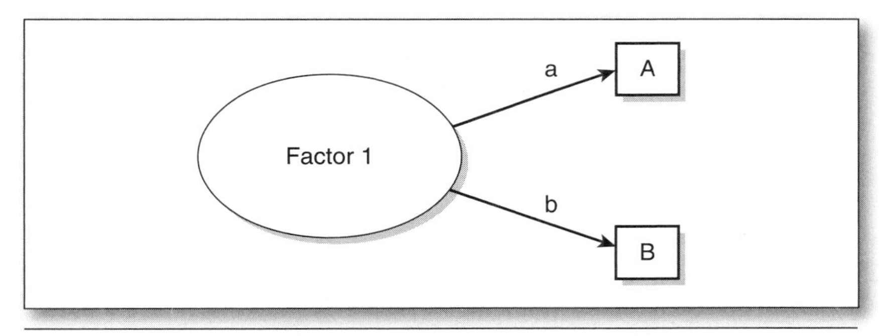

Figure 6.2 A simplified single-factor model involving only two terms

Operations performed on the whole correlation matrix simultaneously do this for each possible pairing of items. Rather than ending with a single residual correlation, one computes an entire matrix of residual correlations (called, appropriately, a residual matrix), each representing the amount of covariation between a particular pair of items that exists above and beyond the covariation that a single latent variable could explain.

Subsequent Factors

It is now possible to operate on this residual matrix in the same way the original correlation matrix was treated, extracting a second factor corresponding to a new latent variable. Once again, correlations between the items and that second latent variable (Factor 2 in Figure 6.3) can be computed, and based

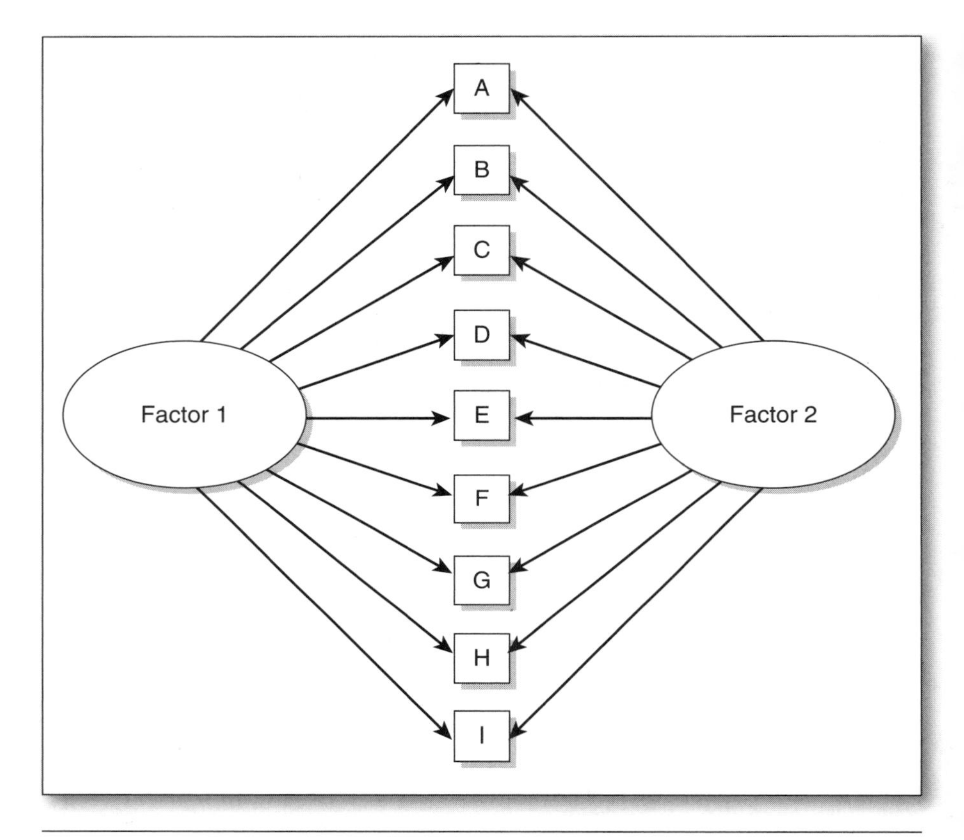

Figure 6.3 A two-factor model

on those correlations, a matrix of proposed correlations can be generated. Those proposed correlations represent the extent of correlation that should remain among items after having taken the second factor into consideration. If the second factor captured all the covariation left over after extracting the first factor, then these projected values should be comparable to the values that were in the residual matrix mentioned above. If not, further factors may be needed to account for the remaining covariation not yet ascribed to a factor.

This process can proceed, with each successive factor being extracted from the residual matrix that resulted from the preceding iteration, until a matrix is achieved that contains only acceptably small residual correlations. At this point, one can determine that essentially all the important covariation has been accounted for and that no further factors are needed. It is possible to continue the process until a residual matrix consisting entirely of zeros is obtained. This will occur when the number of factors extracted equals the number of items in the factor analysis. Stated differently, a set of k factors will always be able to explain all the covariation among a set of k items.

Deciding How Many Factors to Extract

Determining how many factors to extract can be a knotty issue (e.g., Zwick & Velicer, 1986). Of course, a major motivation for conducting factor analysis is to move from a large set of variables (the items) to a smaller set (the factors) that does a reasonable job of capturing the original information (i.e., to condense information). Determining what is a "reasonable job" can be approached in several ways.

Some factor analytic methods, such as those based on maximum likelihood estimation and confirmatory factor analytic procedures (which we will discuss subsequently) based on structural equation modeling approaches, use a *statistical criterion.* In this context, the term *statistical criterion* refers to the fact that inferential methods are used to determine whether the likelihood of a particular result is sufficiently small to rule out its chance occurrence. This amounts to performing a test to see if, after extracting each successive factor, the remaining residuals contain an amount of covariation statistically greater than zero. If they do, the process continues until that is no longer the case. The reliance on a statistical criterion rather than a subjective judgment is an appealing feature of these approaches. However, in scale development, it may not correspond to the goal at hand, which is to identify a small set of factors that can account for the important covariation among the items. Statistically based methods seek an *exhaustive* account of the factors underlying a set of items. If some source of covariation exists that has not been accounted for by any of the factors yet extracted, such programs push further. What the scale developer often is after is a *parsimonious* account of the factors. That is, in the course of scale development, we often want to know about the few, most influential, sources of variation underlying a set of items, not every possible source we can ferret out. When developing a scale, one typically generates a longer list of items than are expected to find their way into the final instrument. Items that do not contribute to the major identifiable factors may end up being trimmed. Our goal is to identify relatively few items that are strongly related to a small number of latent variables. Although a skilled data analyst can achieve this goal by means of factor analytic methods using a statistical criterion, a less experienced investigator, paradoxically, might do better using other more subjective but potentially less cryptic guidelines.

These relatively subjective guidelines are often based on the proportion of total variance among the original items that a group of factors can explain.

This is essentially the same basis used by the statistically based methods. In the case of nonstatistical (i.e., not based on likelihood) criteria, however, the *data analyst* assesses the amount of information each successive factor contains and judges when a point of diminishing returns has been reached. This is roughly analogous to interpreting correlations (such as reliability coefficients) on the basis of their magnitude, a subjective criterion, rather than on their p value, a statistical criterion. Two widely used nonstatistical guidelines for judging when enough factors have been extracted are the eigenvalue rule (Kaiser, 1960) and the scree test (Cattell, 1966).

An *eigenvalue* represents the amount of information captured by a factor. For certain types of factor analytic methods (e.g., principal components analysis, discussed later in this chapter), the total amount of information in a set of items is equal to the number of items. Thus, in an analysis of 25 items, there would be 25 units of information. Each factor's eigenvalue corresponds with some portion of those units. For example, in the case of a 25-item analysis, a factor with an eigenvalue of 5.0 would account for 20% (5/25) of the total information; one with an eigenvalue of 2.5 would account for 10%, and so on. A consequence of this relationship between how information is quantified and the number of items in the analysis is that an eigenvalue of 1.0 corresponds to $1/k$ of the total variance among a set of items. Stated differently, a factor (assuming principal components analysis) that achieves an eigenvalue of 1.0 contains the same proportion of total information as does the typical single item. Consequently, if a goal of factor analysis is to arrive at a smaller number of variables that substantially capture the information contained in the original set of variables, the factors should be more information-laden than the original items. Accordingly, the *eigenvalue rule* (Kaiser, 1960) asserts that factors with eigenvalues less than 1.0 (and, thus, containing *less* information than the average item) should *not* be retained. Although the rationale for excluding such factors makes sense, what about factors that are only slightly above 1.0? Does a factor that explains 1% more information than the typical item really offer the sort of condensation of information we are after? Oftentimes, the answer is no, suggesting that the eigenvalue rule may be too generous a basis for retaining factors. I believe this is generally the case in scale development based on classical methods.

The *scree test* (Cattell, 1966) is also based on eigenvalues but uses their relative rather than absolute values as a criterion. It is based on a plot of the eigenvalues associated with successive factors. Because each factor, after the first, is extracted from a matrix that is a residual of the previous factor's extraction (as described earlier), the amount of information in each successive factor is less than in its predecessors. Cattell suggested that the "right" number of factors can be determined by looking at the drop in amount of information (and, thus, in eigenvalue magnitude) across successive factors. When plotted, this information will have a shape characterized by a predominantly vertical portion on the left (representing large eigenvalues) transitioning to a relatively horizontal portion on the right (corresponding with small eigenvalues). He regarded the factors corresponding with the right-side, horizontal portion of the plot as expendable. In lay terms, *scree* describes the rubble that collects on the ground following a landslide. This term, then, implies that the vertical portion of the plot is where the substantial factors are located while the horizontal portion is the scree, or rubble, that should be discarded. Ideally, the progression of factors will have a point at which the information drops off suddenly, with an abrupt transition from vertical to horizontal and a clear "elbow" (see Figure 6.4). Cattell's criterion calls for retaining those factors that lie above the elbow of the plot. Sometimes, the transition is not abrupt but gradual (see Figure 6.5), with a gentle curve made up of several factors lying between the vertical and horizontal regions of the plot. In such cases, applying Cattell's scree test can be tricky and involves even greater reliance on subjective criteria, such as factor interpretability. A factor is considered interpretable to the

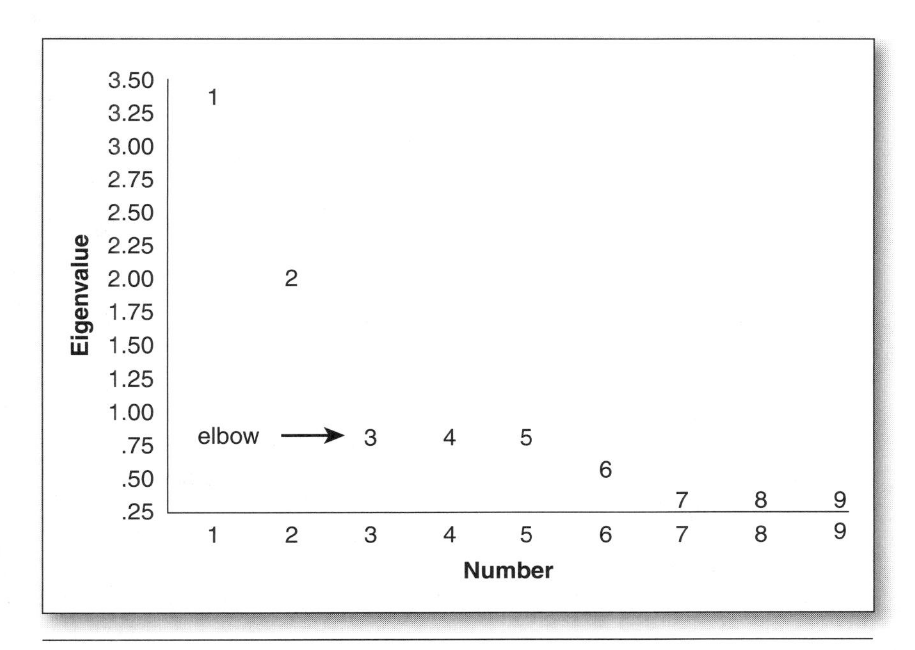

Figure 6.4 A scree plot with a distinct elbow

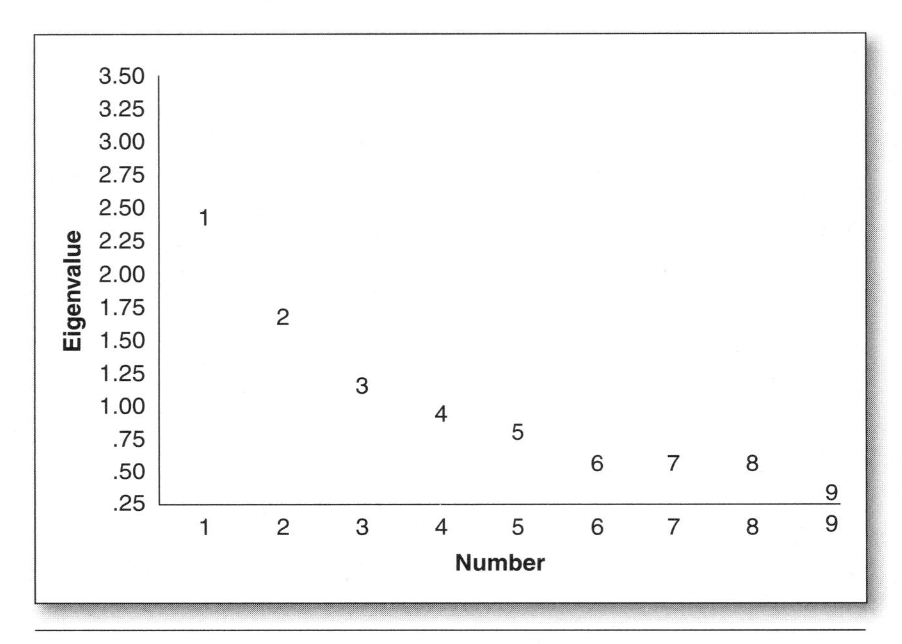

Figure 6.5 A scree plot without a distinct elbow

extent that the items associated with it appear similar to one another and make theoretical and logical sense as indicators of a coherent construct.

Another statistical criterion that is gaining in popularity is based on *parallel analysis* (e.g., Hayton, Allen, & Scarpello, 2004). The logic underlying this approach is that the magnitude of the eigenvalue for the last retained factor should exceed an eigenvalue obtained from random data under otherwise comparable conditions. In other words, in a real study involving the factor analysis of, say, 20 variables measured on 500 people, the eigenvalue of any retained factor should be greater than the corresponding eigenvalue obtained from randomly generated data arranged to represent 20 variables measured on 500 people. Parallel analysis routines, which are available on the Internet as user-developed macros for a variety of statistical packages, generate a large number of distributions of the size corresponding to the actual data set. Thus, in this example, the algorithm would generate many (the user can typically specify how many; 1,000 is often the default) random data sets of 20 variables and 500 subjects. Eigenvalues are extracted for each of these artificial data sets, and a distribution of the eigenvalues is created within the program for Factor I, Factor II, and so forth. For each of these distributions, a representative value (e.g., the median) is identified. Graphical output is displayed, as shown in

Figure 6.6. This figure has two lines connecting points corresponding to eigenvalues for successive factors. The more-or-less straight line descending gradually from left to right represents the median eigenvalues (across the numerous, computer-generated data sets) for essentially random data in which the variables are not related to common underlying factors. The other line in the figure is the actual scree plot, based on the real data in which one is interested. The number of factors to retain is indicated by where the tracings for actual and random data cross. The magnitudes of eigenvalues that lie above the line representing the random data are greater than would be expected by chance alone (i.e., they are statistically significant). Those below the random-data line are not significant (i.e., they can be expected to occur purely by chance). In the example shown in the accompanying figure, the first two factors from the actual data achieved eigenvalues greater than those for the corresponding factors based on the randomly generated data. Thus, only those first two factors meet the retention criterion. Note that the random data yield a number of factors with eigenvalues greater than 1.0, demonstrating the inadequacy of the eigenvalue rule. Examination of the real-data line shows that two factors

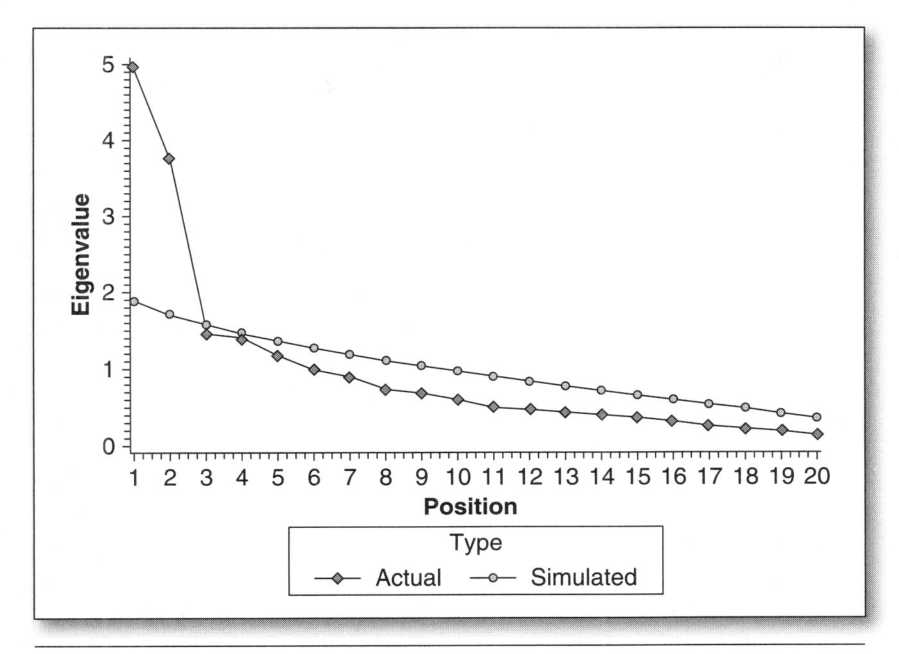

Figure 6.6 A plot from parallel analysis showing magnitudes of eigenvalues for successive factors from random data and actual data

would also be retained based on the scree test criterion of retaining factors above the "elbow" of the plot (which corresponds with the third factor).

Because macros for conducting parallel analysis in various statistical packages (including SAS and SPSS) are readily available via the Internet without restrictions on their use and at no charge, applying this method as a means of selecting the number of factors to retain has become common practice. In my experience, the guidance it provides is consistent with the judgments I would make based on subjective criteria such as the scree plot. It has the advantage, however, of its grounding in a statistical criterion and, thus, has wider acceptability than subjective methods. Accordingly, I recommend using parallel analysis as the primary basis for determining how many items to extract.

Circumstances may arise when substantive considerations (such as the interpretability of resulting factors) support the retention of more or fewer factors than a method such as parallel analysis indicates. When this occurs, I suggest presenting the results of the method first used (e.g., parallel analysis) together with a rationale for deciding to depart from the guidance that method provides. If the substantive issues are compelling, there should be little difficulty in convincing readers and reviewers that an appropriate decision has been reached. On the other hand, if the arguments marshaled in support of ignoring the initial results are not persuasive, then dismissing those results may not be warranted.

Rotating Factors

The purpose of factor extraction is merely to determine the appropriate number of factors to examine. Putting information into the most understandable form is not its intent. The raw, unrotated factors are rather meaningless mathematical abstractions. As a rough analogy, imagine that I have been asked to describe the height of all the people in a room. I decide to do this by arbitrarily selecting a person, Joe, at random, measuring Joe's height, and describing everyone else as so many inches taller or shorter than the reference individual. So, one person might be "Joe plus 3 inches" and another, "Joe minus 2 inches." In such an instance, all the information about height is available in my presentation of the data but it has not been organized in the most informative way. It would be easier for people to interpret my data if I transformed them to a more readily understandable form, such as the height of each individual in the room, expressed in feet and inches. Factor rotation is analogous to this transformation in that it presents data already available in a way that is easier to understand.

Before trying to interpret factors—to ascertain what the constructs or latent variables corresponding with factors are, based on the items identified with each factor—it is usually necessary to perform a factor rotation. Factor rotation increases interpretability by identifying clusters of variables that can be characterized predominantly in terms of a single latent variable (i.e., items that are similar in that they all have a strong association with, and thus are largely determined by, only one and the same factor). Rotation and the greater interpretability that results from it are accomplished not by changing the items or the relationships among them but by selecting vantage points from which to describe them.

The notion that the same set of information can be made more or less meaningful based on the vantage point from which it is inspected can be difficult to grasp. In the following sections, I will provide several examples intended to clarify that idea.

The patterns of intercorrelations among a set of items are analogous to physical locations in space. The more strongly two items are correlated, the closer two markers representing those items could be placed to each other. If we did this for many items, the physical locations of their markers would take on a pattern representing the patterns of correlations among the variables. (This is easiest to visualize if we limit ourselves to two dimensions.) Imagining physical objects whose locations are determined by underlying rules is, thus, another way of thinking about items whose associations are determined by underlying causal variables.

Rotation Analogy 1

How does rotation allow us to see a pattern among variables that was always there but was not apparent? As an analogy, imagine a well-organized array of objects, such as a set of pillars arranged in a series of orderly, parallel rows. It is possible to stand in certain locations and choose certain angles of view such that the arrangement of objects in orderly rows and columns is completely obscured. Changing one's vantage point, however, can reveal the underlying order. Consider how an arrangement of four rows of 10 pillars might appear when viewed from different perspectives. If the line of sight does not reveal any of the natural axes of linear arrangements, the objects may appear to be placed randomly. Viewing from a higher vantage, stepping a few feet right or left, or merely shifting one's gaze laterally can result in the line of sight revealing an alignment of objects and showing their orderliness. Figures 6.7 and 6.8 illustrate this point. These figures were created by plotting locations for the pillars and, literally, altering the vantage point from which the array is viewed. Thus, both figures represent the very same configuration of pillars and differ only in how they are viewed. In the first of these figures, it is difficult at best to discern any organization to the arrangement of pillars. The line of sight

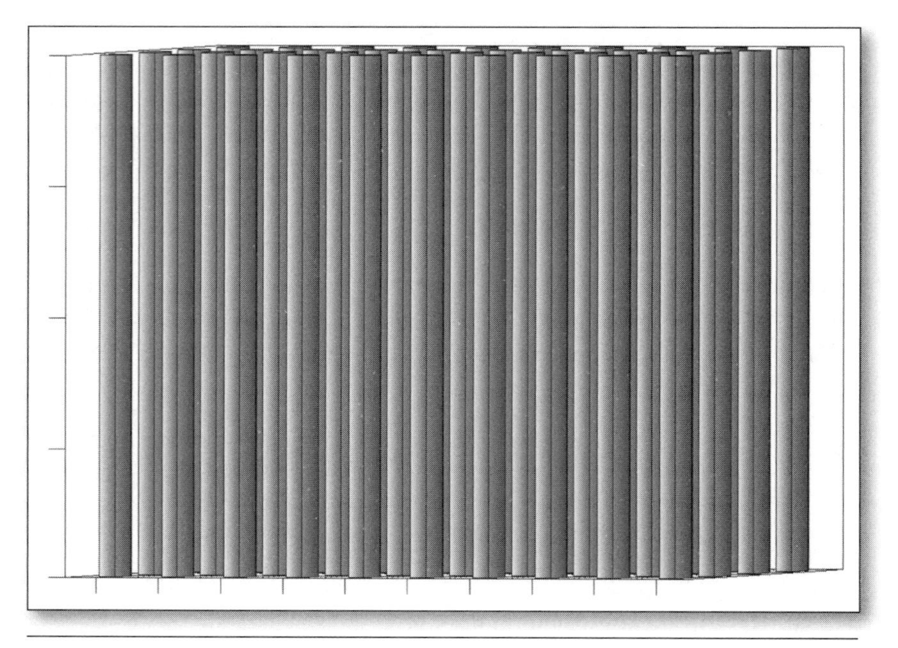

Figure 6.7 An orderly arrangement of pillars viewed from a perspective that obscures that orderliness

through spaces between pillars falls on other pillars farther back, making the grouping appear as a disorganized mass of objects. The second of these figures represents an alternative vantage point. Each pillar, Figure 6.8 reveals, shares a row (and a column) with other pillars. Thus, all the pillars in a given row have something in common—they possess a shared attribute (membership in the same row) that had not been evident from the earlier vantage point. Merely changing our point of view made something about the nature of the objects apparent that was obscured in our initial perspective. Factor rotation is analogous in that it attempts to provide a "vantage point" from which the data's organizational structure—the ways in which items share certain characteristics—becomes apparent.

Rotation Analogy 2

It is worth noting that, with the right number of perpendicular reference lines, one can locate objects no matter how those reference lines are oriented. A two-dimensional example such as a large, empty football field can be used as an illustration. To make this mundane landscape a bit more interesting, let

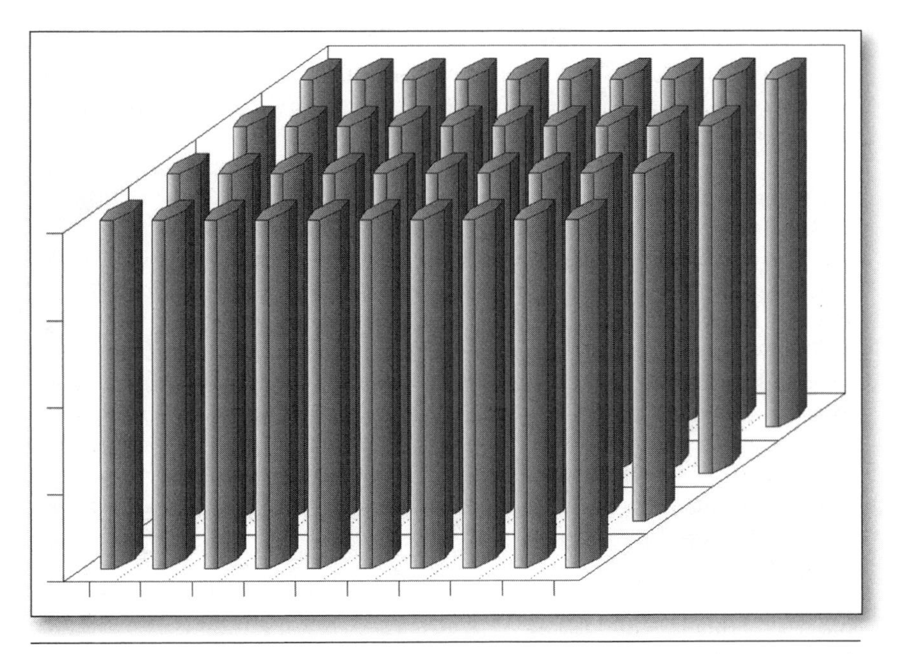

Figure 6.8 An orderly arrangement of pillars viewed from a perspective that reveals that orderliness

us assume that a coach is using a video of a game, shot from an overhead camera, to explain a play to team members. Further, assume that the coach has frozen the image at an especially interesting moment during the game to make a particular point and that this frozen image is projected onto a large computer screen that allows the coach to draw on top of that image. The coach could draw a straight line through the image of the football field at any orientation and then position a second line perpendicular to the first. For example, he or she might choose one line running exactly north and south and another running precisely east and west, which might let him or her describe any location on the field using global positioning coordinates based on north-south latitude and east-west longitude. With those two lines, the coach could specify the location of the player in possession of the ball: "Using the global positioning coordinates for the center of the field as a reference, the ball carrier is at the intersection of a point defined by progressing due northward from the center of the field 60 feet and then proceeding due west an additional 32 feet." This would pinpoint the football at a specific location. The coach could specify the same location using appropriately modified instructions based on any two perpendicular lines drawn through the football field. So, the orientation of the lines is arbitrary with respect to their adequacy of describing a specific location. Any set of two perpendicular lines has the same informational utility in locating a specific spot as any other set.

Alternatively, rather than using the north-south and east-west axes, the coach could use the sidelines and goal lines already drawn on the field as the requisite perpendicular axes. Like any other set of perpendicular lines, these could be used as references for specifying the location of any point on the field. They have the advantage, however, of making use of a meaningful feature of the field—its organization as a grid defined by perpendicular lines (i.e., yard lines and sidelines). Thus, telling someone that the ball carrier was 4 feet inside the 30-yard line and about 68 feet from the home team's sideline might be an easier means of locating that player's position than using north-south and east-west reference lines. The point is that, while either set of reference axes *can* unambiguously locate the ball carrier, the set based on the natural properties of the football field seems considerably more appropriate and comprehensible (see Figure 6.9).

Now that we have seen how vectors or reference lines can be used to specify a spatial location, we should examine how items relate to analogous reference lines. It can be difficult, however, to switch between thinking about items and

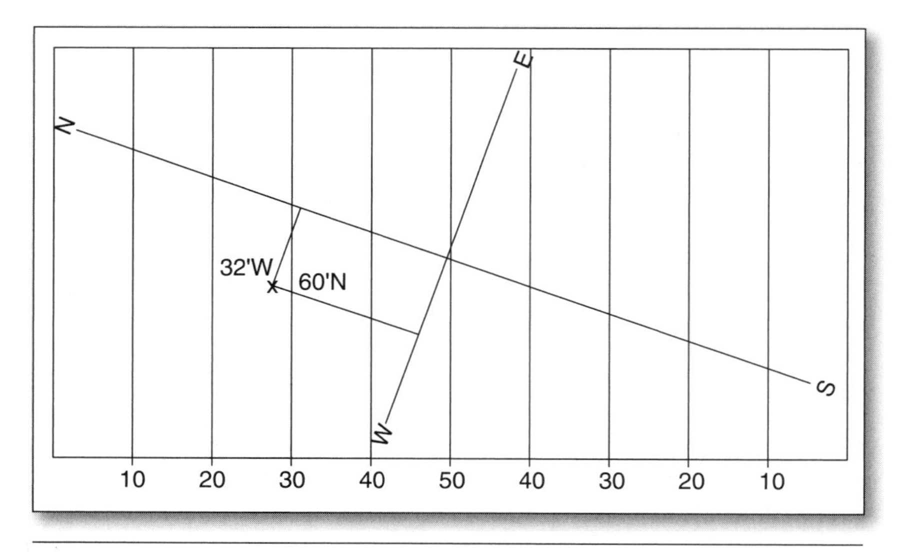

Figure 6.9 A depiction of a football field marked with yard lines, sidelines, and north-south and east-west axes

their content on one hand and thinking about abstract vectors specifying locations on the other. Examples linking the more familiar items with their spatial representations along vectors often must be contrived because the things we measure as social and behavioral scientists do not typically lend themselves to a direct translation in spatial terms. To bridge this gap, I will present a fabricated example based on items that we can link to physical locations.

Rotation Analogy 3

Extending our football example, we will assume a study that uses questionnaire items as a means of describing the location of a football player (the one with the ball) on a field. More specifically, let us assume that a group of respondents is asked to view a video of a football game and that, at some point, the image is frozen and the respondents are asked to complete a series of eight items using a Likert scale with responses ranging from 1 = *Strongly Disagree* to 6 = *Strongly Agree*. The items are as follows:

A. The ball carrier is within easy scoring position.
B. The ball carrier may soon be forced out of bounds.
C. The ball carrier may be forced into his own end zone.
D. The ball carrier has plenty of room to either side.
E. If the ball carrier were tackled right now, someone on the sidelines might be in danger.
F. The ball carrier has a long way to go before crossing the goal line.
G. The ball carrier is within field goal range.
H. The ball carrier needs to be careful not to step over the sideline.

Before continuing with this example, let us take a moment to examine more formally how rotation is accomplished. When there are two factors, rotation can be achieved graphically. One can obtain a scatterplot that represents the strength of correlations between pairs of items by the proximity of those items and the loadings of those items on arbitrary reference vectors corresponding with two unrotated factors. One can perform a graphical rotation by literally tracing those axes and then rotating the tracing until the lines fall along item groupings. Then, the reference axes are used to generate coordinates defining the exact position of each represented item in terms of its location relative to the lengths along the two axes. These coordinates can then be converted into loadings for each item on the two factors. But automated nongraphical rotation methods based on mathematical criteria are far more widely used. For example, methods such as Varimax seek to maximize the variance of the squared loadings for each item. Loadings are correlations between each item and each factor based on the orientation of the vectors (i.e., reference lines) defining the factors. So, depending on how the vectors are oriented, the loadings will vary, just as the values specifying location of a football player on the field or the coordinates describing the position of a dot on a scatterplot will vary depending on which reference axes are used. This variance will be greatest when some squared loadings are relatively large and others are relatively small. Thus, satisfying the criterion of maximizing the variance of the squared loadings has the effect of producing a maximally uneven set of loadings for each item. Finding an orientation for the reference axes that will give us maximally uneven loadings and the largest squared loading variances amounts to finding a potentially informative rotation. What we are seeking, of course, is a specific type of unevenness in which one loading (i.e., the loading of the item on its primary factor) is substantial while all other loadings for that item (i.e., its loadings on secondary factors) are small (ideally, near zero). When this is achieved, we can describe the item as being influenced primarily by the single factor on which it loads substantially—a circumstance called *simple structure.* Describing an item in terms of only its primary factor does a good job of capturing the essential nature of the information that item carries. That is the goal of rotation.

Now, let us go back to our items describing the location of the ball carrier in the frozen image of the football game. I ran an analysis of those eight items (based on fabricated data) for illustrative purposes. The first three eigenvalues were 3.67, 2.78, and .59. Based on the fact that the first two eigenvalues are substantially larger and the third is substantially smaller than 1.0, I extracted two factors. I requested that SAS perform a Varimax rotation on the factors. The SAS factor plots before and after rotation are depicted in Figures 6.10 and 6.11, respectively. These locate the individual items with respect to the two factors based on the loading of each item on those two factors.

SAS prints its factor plots in a typeface that uses equal width for each character and cannot interpolate between lines of text. Consequently, it cannot locate the points relative to one another or to the factor vectors with exact precision. Even with this limitation, however, the plots are quite revealing. In the first (Figure 6.10), the items (represented by bolded letters in shaded boxes for clearer visibility) are clustered in two groups, one comprising items B, D, E, and H and the other, items A, F, C, and G. Proximity corresponds with strength of correlation in this figure. So, any two items that are located near each other are more highly correlated than any pair that are more distant. The presence of two clusters in this plot, therefore, indicates that the items within each cluster correlate substantially with one another and that items in different clusters are largely uncorrelated with one another. Positions relative to the axes are also informative and especially pertinent to the issue of rotation. The

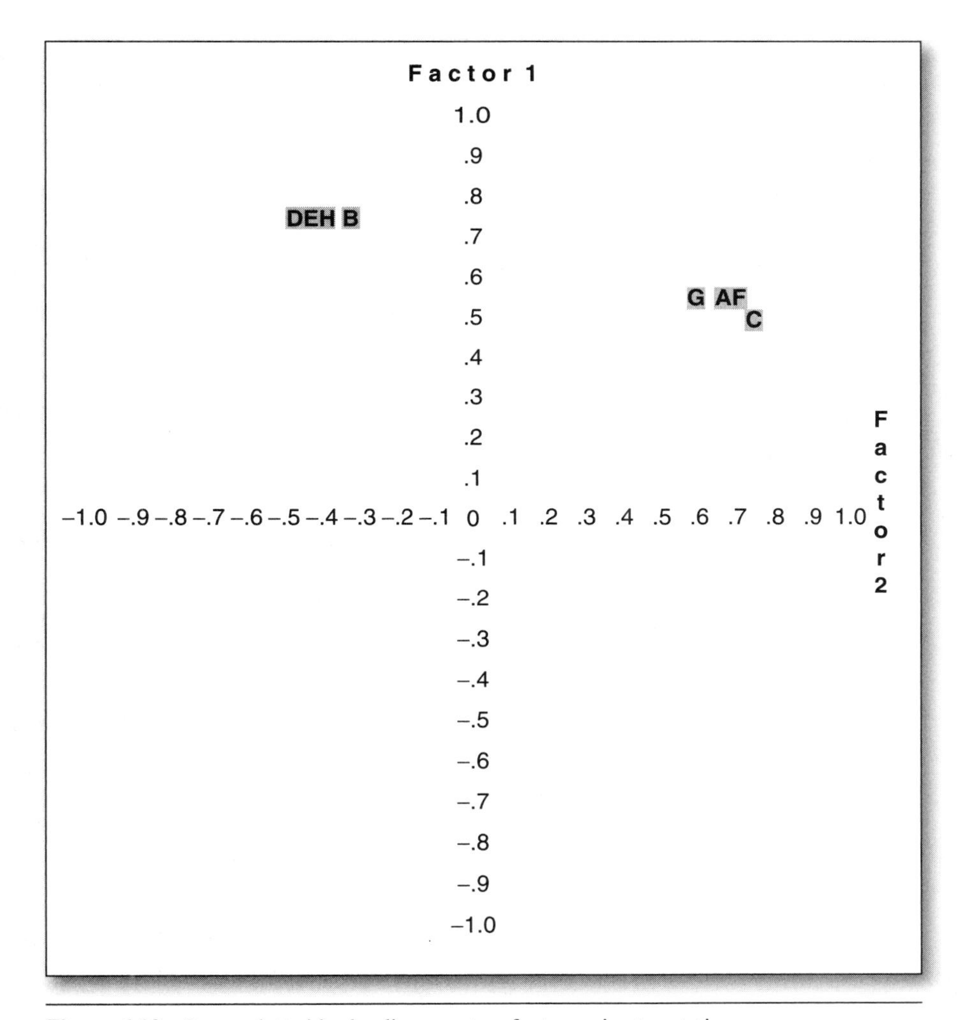

Figure 6.10 Items plotted by loadings on two factors prior to rotation

A-F-C-G item cluster is located roughly equidistant from the two vectors representing the factors. That is, its loadings on each factor are roughly the same magnitude. The other item cluster, while somewhat closer to the vertical than the horizontal vector, is still at some distance from either. The items in that second cluster have higher loadings on Factor 1 but also fairly substantial (negative) loadings on Factor 2. Consequently, it would be hard to say that either cluster of items primarily reflects the level of just one of the factors.

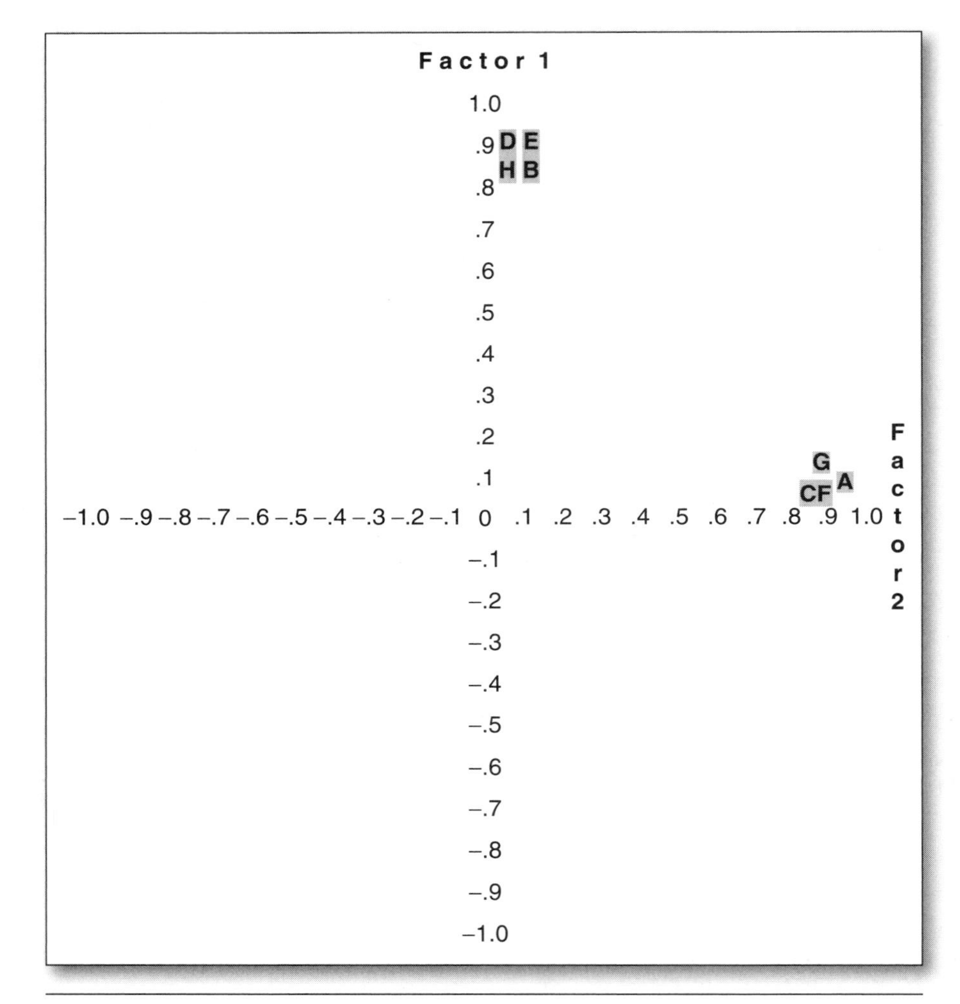

Figure 6.11 Items plotted by loadings on two factors after rotation

In contrast, in the plot made after the factors were rotated, the two clusters are very close to one vector each. The first cluster of items (B-D-E-H) now has high loadings (roughly .8 to .9) on Factor 1 (represented by the vertical axis) and low loadings on Factor 2 (less than .2). The second cluster (A-F-C-G) is contiguous to the Factor 2 (horizontal) axis, and the items collectively represent high values on Factor 2 (roughly in the range of .8 to .9) and low values on Factor 1 (about .1). So, after rotation, we have a much better approximation to simple structure—that is, each item loads primarily on a single factor.

If you look carefully at the relative locations of the two item clusters on the two different plots, you may be able to discern that they are quite similar (within the limitations of SAS's plotting). The A-F-C-G cluster is roughly 90 degrees clockwise of the B-D-E-H cluster in both figures. This is as it should be, because rotation does not alter the relationships among the variables. If the items were drawn on one transparent sheet and the factors on another, you could rotate the sheet with the items from the position shown in Figure 6.10 to the position shown in Figure 6.11 relative to the factors. (The fact that the horizontal and vertical axes have slightly different scales because of the imprecision of how SAS displays the plots may make the distances of the clusters from the origin seem less similar in the two figures than they actually are numerically.) In fact, rotation actually reorients the factors, but rather than displaying the factor axes at odd angles, the post-rotation plot essentially rotates the item locations clockwise rather than rotating the axes counter-clockwise.

Having rotated the factors, it is now possible to interpret each of them by looking at what the items achieving high loadings on each of the factors have in common. The items composing the first factor all concern proximity to sidelines, and those composing the second concern proximity to goal lines. Thus, we might interpret Factor 1 as concerning *Proximity to Sidelines* and Factor 2 as *Proximity to Goal Lines.* Rotation has allowed us to group the items in a way that made it easier to recognize the common attribute shared by the items within each group.

In summary, the most important points about factor rotation that these examples illustrate are as follows: (a) Knowing the correct number of factors (when there is more than a single factor), while essential, does not reveal the nature of those factors. The initial factor extraction process addresses only the number of factors needed, not the nature of those factors in any substantive sense. (b) In the absence of rotation of multiple factors, any meaningful and interpretable pattern among the items may be obscured. The initial solution ignores item content, and the item groupings based on unrotated factors will typically reveal little about the items. The depiction of the pillars as viewed from a disadvantageous and then an advantageous perspective illustrated this point. (c) Any analytic solution comprising the correct number of factors can specify the relationships among the items (or, if we depict them spatially, the locations of the items relative to one another) as well as any other. North-south and east-west axes work just as well as goal-to-goal and sideline-to-sideline axes for locating any position on the field. (d) The purpose of factor rotation is to find a particular orientation for the reference axes that helps us understand items in their simplest terms. That is achieved when most of what an item is about can be explained using as few variables as possible—ideally, only one. (e) The rotated factor pattern can facilitate determining what the items within a factor have in common and thus inferring what the underlying causal factor is that determines how the items are answered. The two sets of items regarding the location of the ball carrier in a football game were distinctly different. One set concerned only proximity to the sidelines, while the other concerned only proximity to the goal lines. We can infer, at least provisionally, that perceptions of proximity to those different borders of the football field determined how respondents evaluated the two sets of items. Of course, additional validation information would be needed to support this inference.

Orthogonal Versus Oblique Rotation

All the examples thus far have been based on reference lines that are perpendicular to one another. This corresponds with factors that are statistically independent of one another (i.e., uncorrelated). Such factors are described as *orthogonal.* Location along one line provides no information regarding information along another when the two are perpendicular. For example, knowing how far north someone is gives no indication of how far west they are, because those two directions are orthogonal to each other. Similarly, knowing how far a player is from the goal line gives no indication of proximity to the sideline.

It is possible, however, to allow factors to be correlated and, consequently, for the axes that represent them graphically to be nonperpendicular. Suppose, for example, that we chose to specify locations on the football field depicted in Figure 6.9 using a line running from end zone to end zone as one axis and a line running due east and west as the other. Moving in a westerly direction also means moving toward one of the end zones. That is, one cannot move directly east or west without changing one's distance from the goal lines. The two dimensions are correlated to some degree.

Factor rotation can likewise allow reference axes (and the factors to which they correspond) to be correlated and, thus, not be spatially perpendicular. Such a rotation is referred to as *oblique* rather than orthogonal. Oblique rotation may be useful when the underlying latent variables are believed to correlate somewhat with one another. The goal of simple structure requires items that can be meaningfully classified with respect to only a single category. That is, each item should be "about" only one thing and, thus, load on only one factor. If the variables are somewhat correlated but the factors representing them are forced to be totally independent because of constraints imposed by the factor analytic method, it may be impossible to achieve this goal. That is, more than one factor may be associated with some or all of the items because of the correlation between the factors; we are limited in our ability to approximate simple structure (see Figure 6.12).

If "Conscientiousness" and "Dependability" are truly correlated, to return to the earlier coworker-characteristics example, then an item about one is likely

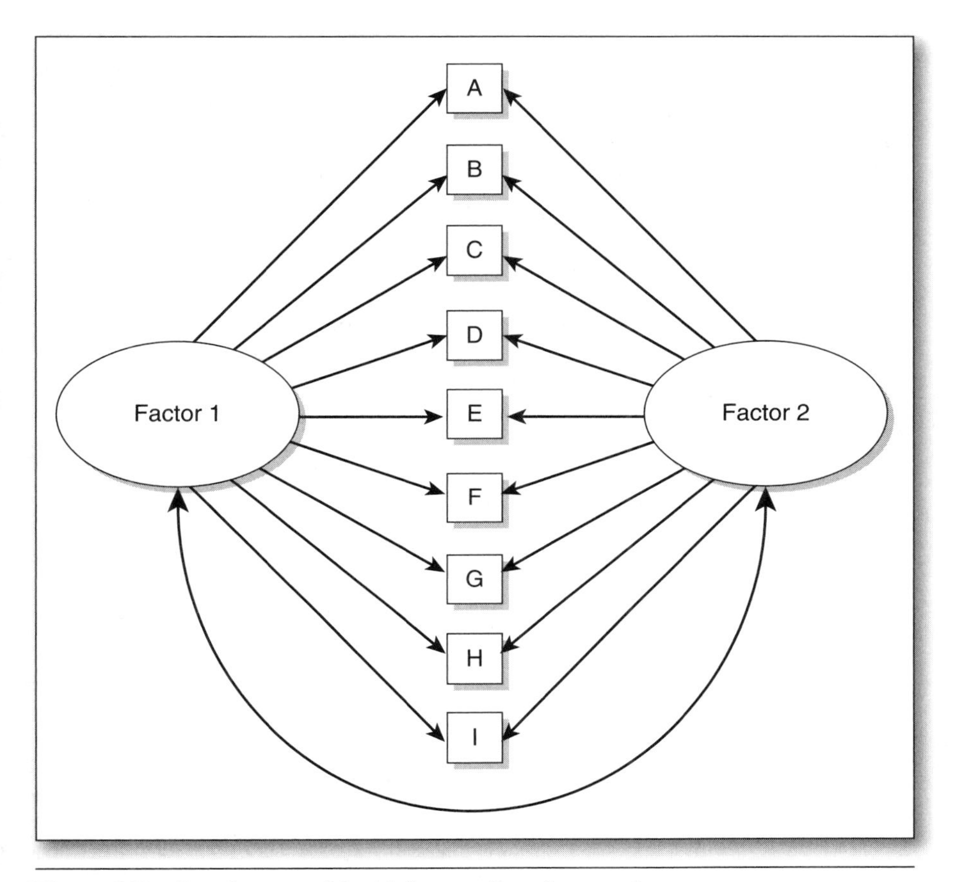

Figure 6.12 Two-factor model with factors allowed to correlate

to share some variance with the other as well. People who are more conscientious will also be somewhat more dependable (to a greater or lesser degree depending on the strength of the correlation between the two attributes). Forcing an orthogonal solution on data reflecting these two variables will make it very hard for an item to load strongly on one variable and weakly on the other. If, however, the two factors are allowed to be somewhat correlated, the circumstance is roughly analogous to the following reasoning: *"Conscientiousness" and "Dependability" are understood to correlate with each other. This fact has been dealt with by allowing the factors to correlate. Now, that aside, to which of these factors is the item in question most strongly related?* If the axes representing the "Conscientiousness" factor and the

"Dependability" factor are not forced to be perpendicular, then as scores increase on one of the variables, some degree of concomitant increase in the other can be accommodated. If reality dictates that an item substantially influenced by "Conscientiousness" is also influenced by "Dependability," allowing the factors representing those constructs to correlate can more accurately accommodate that reality. An item that is strongly related to "Conscientiousness" but mildly related to "Dependability" may be able to load strongly on the first of those factors and weakly on the second if the factors are not constrained to statistical independence. In a sense, the fact that this item has some causal influence arising from "Dependability" is accommodated by allowing the two factors to correlate with each other. This provides an indirect

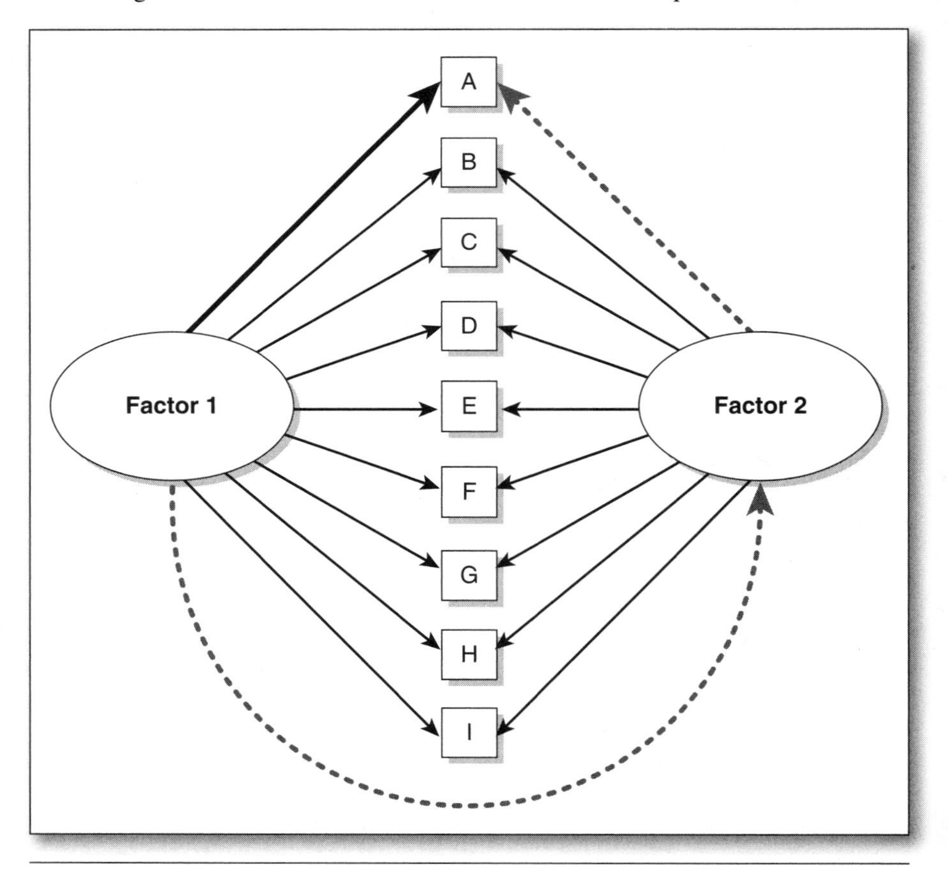

Figure 6.13 Because of the correlations between the factors, Factor 1 influences Item A both directly (as shown by the dark, solid pathway) and indirectly (shown by the lighter, dashed pathway)

causal path from "Dependability" to the item through "Conscientiousness" (see Figure 6.13), thus removing the need for the item to correlate directly with "Dependability" and consequently cross load on both factors.

What is lost when factors are rotated obliquely is the elegance and simplicity of uncorrelated dimensions. One nice feature of uncorrelated factors is that their combined effects are the simple sum of their separate effects. The amount of information in a specific item's value that one factor explains can be added to the information that another factor explains to arrive at the total amount of information explained by the two factors jointly. With oblique factors, this is not the case. Because they are correlated, there is redundancy in the information contained in the two factors. For an item related to both "Conscientiousness" and "Dependability," the amount of variation explained by those two factors together is less than the sum of the parts. Some, perhaps a great deal, of the information accounted for by one factor overlaps with the information accounted for by the other. A simple sum would include the overlapping information twice, which would not accurately reflect the total influence of the factors on that item.

Another complication of correlated factors is the added complexity of the causal relationships between items and factors. When the factors are independent, the only relationship between a factor and a specific item is direct. Changes in the level of the factor will result in changes in the item along a single, direct causal pathway. When factors are correlated, however, this is not the case. If (as described earlier when discussing an item influenced by "Conscientiousness" and "Dependability") both of two hypothetical factors influence Item A, for example, and the factors are correlated, the factors exert an indirect as well as a direct influence on A. That is, Factor 1 can influence Factor 2 and, through Factor 2, indirectly affect Item A. This is in addition to the direct effect of Factor 1 on that item.

Of course, by a parallel process, Factor B also can affect the item not only directly but also indirectly through its relationship to Factor A. The same sort of direct-plus-indirect influence of factors also may apply to all other items. As a result, speaking of the relationship between an item and a factor usually has to be qualified explicitly to either include or exclude such indirect effects. Otherwise, there is ambiguity and, thus, potential confusion.

Choosing Type of Rotation

As a practical matter, the choice between orthogonal versus oblique rotation should depend on one or more considerations. Among these is how one views the concepts the factors represent. If theory strongly suggests correlated concepts, it probably makes sense for the factor analytic approach (specifically, rotation) to follow suit. Thus, if analyzing items related to "Conscientiousness" and "Dependability," allowing factors to correlate would best fit our sense of what these concepts imply. Alternatively, theory might suggest orthogonal factors. "Dependability" and "Fun," for example, may be more independent and, thus, might warrant an orthogonal solution. When theory does not provide strong guidance, as when the scale under development represents concepts not previously studied, the magnitude of the correlations between factors may serve as a guide. Specifically, an oblique rotation can be specified and the resultant correlations among factors examined. If these are quite small (e.g., less than .15), the data analyst might opt for orthogonal rotation. This slightly compromises the approximation of simple structure but results in a simpler model. For example, some items may show secondary loadings (i.e., loadings on a factor other than the one on which they load most strongly) that are slightly increased relative to the oblique solution but still small enough to associate each item unambiguously with only one factor. Thus, loadings of a particular item on three factors rotated to an oblique solution might be .78, .16, and .05. When an orthogonal solution is chosen, the loadings might be .77, .19, and .11. Although the second pattern departs slightly more than the first from simple structure, the item in question can still be linked unambiguously to the first factor. Thus, little has been sacrificed in opting for the simpler (i.e., orthogonal) model in this case. If the factors are more substantially correlated, opting for the oblique solution might yield a substantial improvement in approximation to simple structure. For example, a secondary loading of .40 obtained with an orthogonal rotation might diminish to .15 with an oblique solution. This will not always be the case, however, and only examining the difference between the two rotation methods can definitively indicate how much they differ in approximating simple structure.

A final practical issue concerns the magnitude of the correlation between two factors and how large it needs to be before combining them into one larger factor. There is no simple answer to this question because the relationships of items to factors also need to be considered. Under some circumstances, however, an oblique rotation may reveal that even when two factors are highly correlated, some items have substantial loadings on both. In that case, it might make good sense to extract one fewer factor to see if the two that were highly correlated factors merge into one. It might very well be, for example, that real-world data would support a single factor combining items about conscientiousness and dependability rather than separating them.

INTERPRETING FACTORS

In the example involving "Conscientiousness" and "Dependability" items, we have assumed that we knew a priori exactly what the latent variables were. Often, this is not the case, and then we will rely on the factor analysis to give us clues regarding the nature of those latent variables. This is done by examining the items that most strongly exemplify each factor (i.e., that have the largest loadings on a particular factor). The items with the highest loadings are the ones that are most similar to the latent variable (and, thus, correlate most strongly). Therefore, they can provide a window into the nature of the factor in question. This is most easily done when there are several items that clearly tap one common variable with quite substantial loadings (e.g., greater than .65) on the same factor. Returning to the example of characteristics considered important in a coworker, if "smart," "has a mind like a steel trap," "is well educated," and perhaps one or two other items related to intellectual capacity all loaded substantially on the same factor, with no other items having large loadings on that factor, it would be fairly easy to conclude that "Importance Ascribed to Intellect" or some equivalent description was an apt label for that factor.

Although choosing a label for a factor may seem straightforward in some cases, assigning a name is *not* the same as establishing validity. Whether the item set continues to perform as the assigned name implies will ultimately determine validity. When factors explain relatively little variance and have multiple, seemingly disparate items that load comparably on them, the factor analyst should be especially cautious in interpretation. If the analysis yielded one factor with items that seem dissimilar, it probably is best not to take the factor too seriously as an indicator of a latent variable.

Another point worth remembering at the interpretation stage is that a factor analysis can find the structure accounting for associations only among the items analyzed—it will not necessarily reveal the nature of phenomena per se. A researcher attempting to determine the fundamental dimensions of personality, for example, could not obtain an "Extroversion" factor if no items pertaining to extroversion were included.

Sometimes, inclusion of a specific phrase can create a false appearance of a conceptually meaningful factor. When some statements are worded in the first person and others are not, for example, that may account for the pattern of associations observed. As an illustration, consider the following hypothetical items:

1. I like apples.
2. Oranges taste good.
3. I prefer apples to some other fruits.
4. There are many people who like oranges.
5. I enjoy an apple every now and then.
6. Oranges generally have a pleasant fragrance.
7. I find the crispness of apples appealing.
8. A fresh orange can be a tasty treat.

Were the odd items to load on one factor and the even items on a second factor, we would not know if the "I" wording of the odd items was the cause of the two factors or if people were expressing differential attitudes toward the two types of fruit mentioned. Both explanations are plausible but confounded. This is a case where we may or may not be comparing apples to oranges.

PRINCIPAL COMPONENTS VERSUS COMMON FACTORS

There are two broad classes of data analytic techniques that some authors regard as fundamentally the same but others view as fundamentally different. These are factor analysis and principal components analysis. The term *factor analysis* is sometimes used to embrace both techniques and at other times to describe one as opposed to the other. The terms *common factors* and *components* are often used as a less ambiguous way of referring specifically to the composites arising from factor analysis and principal components analysis, respectively. There is a basis for asserting both the similarity and the dissimilarity of these methods.

Principal components analysis yields one or more composite variables that capture much of the information originally contained in a larger set of items. The components, moreover, are defined as weighted sums of the original items. That is, principal components are linear transformations of the original variables. They are grounded in actual data and are derived from the actual items. They are merely a reorganization of the information in the actual items.

Common factor analysis also yields one or more composite variables that capture much of the information originally contained in a larger set of items. However, these composites represent hypothetical variables. Because they are hypothetical, all we can obtain are estimates of these variables. A common factor is an idealized, imaginary construct that presumably causes the items to be answered as they are; the nature of the construct is inferred by examining how it affects certain items.

Same or Different?

The above descriptions highlight some differences between components and factors. One of those differences is that factors represent idealized, hypothetical variables we estimate whereas components are alternative forms of the original items with their information combined. The idea behind extracting common factors is that we can remove variance from each item that is not shared with any of the other items. From the perspective of factor analysis, as was the case with reliability, unshared variation is essentially error. Thus, the combinations we arrive at in extracting common factors are estimates of hypothetical, error-free, underlying variables. It is in this sense that common factors are idealized—they are estimates of what an error-free variable determining a set of items might look like. Furthermore, factors *determine* how items are answered, whereas components are *defined by* how items are answered. Thus, in principal components analysis, the components are end products of the items, and the actual scores obtained on items determine the nature of the components. In common factor analysis, however, we invoke the concept of an idealized hypothetical variable that is the cause of the item scores. A factor is an estimate of that hypothetical variable and represents a cause, not an effect, of item scores.

What about the similarities? There are several.

First, the computational difference between the two is minor. Remember, in common factor analysis, the goal is to estimate an idealized, error-free variable. But we must generate this estimate from the actual data. As we have noted, factor analytic methods generally are based on a correlation matrix representing all the associations among the items to be factored. Back in Chapter 3, I pointed out that all the off-diagonal values in a covariance or correlation matrix represent only shared, or communal, variance. As I pointed out then, a correlation matrix is simply a standardized version of a variance-covariance matrix. The correlations themselves are standardized covariances, and the unities are standardized item variances. Each standardized item variance represents all the variability, both shared and unique, that an item manifests. To create an idealized, error-free variable, the unique portion of variance contained in the item variances along the main diagonal of the correlation matrix must be purged. More specifically, each unity must be replaced by a *communality estimate,* a value less than 1.0 that approximates only a given variable's shared variation with other variables included in the factor analysis. For example, if we estimated that a particular variable shared 45% of its total variation with other items in the correlation matrix, we would assign it a communality estimate of .45 and place that value in the matrix, replacing the 1.0 that represented the item's total variance. We would do this for every variable, replacing each unity with a communality estimate. (Often, communality estimates are obtained by regressing the variable in question on all the remaining variables so as to obtain the squared multiple correlation, R^2, which serves as the estimate.) This purging process creates an altered correlation matrix that is used for extracting common factors rather than components, as shown in Table 6.1.

Substituting communality estimates for unities is the only computational difference separating the extraction of common factors from the extraction of principal components.

1.0	.70	.83	.48	.65
.70	1.0	.65	.33	.18
.83	.65	1.0	.26	.23
.48	.33	.26	1.0	.30
.65	.18	.23	.30	1.0

.45	.70	.83	.48	.65
.70	.52	.65	.33	.18
.83	.65	.62	.26	.23
.48	.33	.26	.48	.30
.65	.18	.23	.30	.58

Table 6.1 Correlation Matrices for Principal Components Analysis and Common Factor Analysis

Note: The correlation matrix on the left, which is used for principal components analysis, retains unities in the main diagonal. The correlation matrix on the right, which is used for common factor analysis, has communality estimates, rather than unities, along the main diagonal.

What about the cause versus effect issue? We obtain both factors and components by analyzing scores on observed items. As the explanation of communality estimates demonstrates, empirical relationships among the items ultimately are the foundation for common factors. This, of course, is also true for components. So, computationally, both are grounded in empirical data. Furthermore, most data analysts conceptualize both components and common factors as ways of understanding the variables underlying a set of items. That is, both components and factors are customarily thought of as revealing the cause for observed scores on a set of items. Components analysis and factor analysis, in fact, are often used interchangeably; under most circumstances in which items have something meaningful in common, the different methods support the same conclusions. In many factor analytic routines (such as PROC FACTOR in the SAS statistical package), principal components analysis is the default. That is, unities are retained in the correlation matrix unless communalities are specified. So, while there are both technical similarities and differences between the two, the distinctions between them are often overlooked with few if any adverse consequences.

One important difference, however, is the nature of the variance explained by components versus factors. The former account for a specified portion of the *total* variance among the original variables, whereas the latter account for the *shared* or *common* variance. Reducing the diagonal values of the correlation matrix, as one does when extracting common factors, reduces both the numerator and the denominator of the ratio expressing proportion of variance. But it reduces the denominator to a greater degree because of the specific calculations involved in computing the relevant variances. As a result, the *proportion of variance* explained by a set of comparable components and factors will not be equal or conceptually equivalent. Factors will explain a larger proportion of a more restricted variance (i.e., shared variance), while components will explain a smaller proportion of total variance. When discussing factor analytic results and reporting the proportion of variance explained by the factors, it is critical to be clear about what type of analysis (components or common factors) and, thus, what type of variance (shared or total) is being explained.

Another difference worth noting is that, in some statistical packages, some of the output obtained from the extraction of common factors will appear nonsensical. In both types of analysis, the cumulative amount of variance explained will mount as each successive factor/component is extracted. With common factors, this proportion often exceeds 1.0 at some point, continues to rise as successive factors are considered, and then, as if by magic, returns to a value of precisely 1.0 as the *k*th (i.e., last possible) factor is extracted. Although this looks strange, it is simply an artifact of the computation method and can be ignored. If the data analyst has used reasonable criteria for deciding how many factors to extract, the number chosen will typically precede the point in the extraction sequence where this anomaly arises. It is possible, however, for the selected number of factors to explain virtually all (i.e., 100%) the shared variance among the original items.

CONFIRMATORY FACTOR ANALYSIS

Another bifurcation of factor analytic methods differentiates between exploratory and confirmatory techniques. These terms originally referred to the intent of the data analyst rather than the computational method. Thus, the same analysis might be used on the same set of items either to determine what their underlying structure is (exploratory) or to confirm a particular pattern of relationships predicted on the basis of theory or previous analytic results (confirmatory). With increasing frequency, these terms are now used to differentiate between different types of analytic tools rather than different research objectives. When people use the term *confirmatory factor analysis*, they are often talking about methods based on structural equation modeling (SEM). Although these methods should be used in a confirmatory rather than exploratory fashion, standard factor analytic techniques can be used for either. Thus, "confirmatory" does not necessarily equate to SEM-based.

The SEM-based methods, however, offer some real benefits over traditional factor analytic methods in certain situations. These benefits arise because the SEM models are extremely flexible. Conditions that are assumed by traditional factoring methods, such as independence of the item error terms from one another, can be selectively altered in SEM-based approaches. Also, traditional methods for the most part constrain the data analyst either to allowing factors to correlate with one another or requiring that all be independent of one another. SEM-based approaches can mix correlated and uncorrelated factors if theory indicates such a model applies.

SEM-based approaches also can provide a statistical criterion for evaluating how well the real data fit the specified model. Used judiciously, this can be an asset. Sometimes, however, it can lead to over-factoring. Extracting more factors often improves a model's fit. Applying a strictly statistical criterion may obscure the fact that some statistically significant factors may account for uninterestingly small proportions of variance. Especially in the early stages of instrument development, this may be contrary to the goals of the investigator who is concerned with finding the smallest number of most information-laden factors rather than accounting for the maximum amount of variance possible.

Another double-edged sword in SEM-based approaches is the common practice of testing alternative models and comparing how they fit the data. Again, used prudently, this can be a valuable tool. Conversely, used carelessly, it can result in model specifications that may make little theoretical sense but result in statistically better model fit. As an example, removing the constraint that item errors be uncorrelated with one another might yield quite small values for the correlations, but the model might still statistically outperform a constrained model. One researcher may decide to ignore the small correlations in favor of a simpler model, while another is persuaded by a statistical criterion to reject the more parsimonious alternative. As another example, a model that separates two distinct but highly correlated factors (perhaps like "Conscientiousness" and "Dependability") might fit better than one that combines the two. If the correlation between them is very high, the decision to keep them separate may seem arbitrary. For example, a correlation of, say, .85 between two indicators of the same construct would usually be considered good evidence of their equivalence. But a model that specified separate factors that correlated with each other at .85 might fit the data better than a model that combined the two into a single factor.

These comments are not intended to suggest that SEM-based approaches to confirmatory factor analysis are bad. The advent of these methods has made enormous contributions to understanding a variety of measurement issues. I am suggesting, however, that the inherent flexibility of these approaches creates more opportunities for making poor decisions, particularly when the data analyst is not familiar with these methods. With the possible exception of principal components analysis (in which factors are linear combinations of the items), no factoring method produces a uniquely correct solution. These methods merely produce plausible solutions, of which there may be many. There is no guarantee that a more complex model that statistically outperforms a simpler alternative is a more accurate reflection of reality. It may or it may not be. With all factor analytic approaches, common sense is needed to make the best decisions. The analyses are merely guides to the decision-making process and evidence in support of those decisions. They should not, in my judgment, entirely replace investigator decision making. Also, it is important that the basis for decisions, statistical or otherwise, be described accurately in published reports of confirmatory factor analysis.

One last note on this subject: Researchers in some areas of inquiry (e.g., personality research) consider obtaining consistent results from traditional factoring methods as stronger confirmatory evidence than demonstrating good model fit according to a statistical criterion. For example, Saucier and Goldberg (1996) state that, "because exploratory factor analysis provides a more rigorous replication test than confirmatory analysis, the former technique may often be preferred to the latter" (p. 35). The reasoning is that if data from different samples of individuals on different occasions produce essentially identical factor analytic results using exploratory approaches, the likelihood of those results being a recurring quirk is quite small. Remember that, in SEM-based approaches to this same situation, the data analyst specifies the anticipated relationships among variables and the computer program determines if such a model can be reconciled with the empirical data. In other words, the computer is given a heavy hint as to how things should turn out. In contrast, rediscovering a prior factor structure without recourse to such hints, as may happen with repeated exploratory analyses, can be very persuasive.

USING FACTOR ANALYSIS IN SCALE DEVELOPMENT

The following example should make some of the concepts discussed in this chapter more concrete. Some colleagues and I (DeVellis et al., 1993) developed a scale assessing parents' beliefs about who or what influences their children's health. Although the full scale has 30 items and assesses several aspects of these beliefs, for this presentation, I will discuss only 12 of the items:

A. I have the ability to influence my child's well-being.

B. Whether my child avoids injury is just a matter of luck.

C. Luck plays a big part in determining how healthy my child is.

D. I can do a lot to prevent my child from getting hurt.

E. I can do a lot to prevent my child from getting sick.

F. Whether my child avoids sickness is just a matter of luck.

G. The things I do at home with my child are an important part of my child's well-being.

H. My child's safety depends on me.

I. I can do a lot to help my child stay well.

J. My child's good health is largely a matter of good fortune.

K. I can do a lot to help my child be strong and healthy.

L. Whether my child stays healthy or gets sick is just a matter of fate.

These were administered to 396 parents, and the resultant data were factor analyzed. The first objective of the factor analysis was to determine how many factors there were underlying the items. SAS was used to perform the factor analysis, and a scree plot was requested. A scree plot similar to the type printed by SAS appears below (see Figure 6.14). Note that 12 factors (i.e., as many as the number of items) are plotted; however, 2 of those are located on the initial portion of the plot and the remainder form the scree running along its bottom.

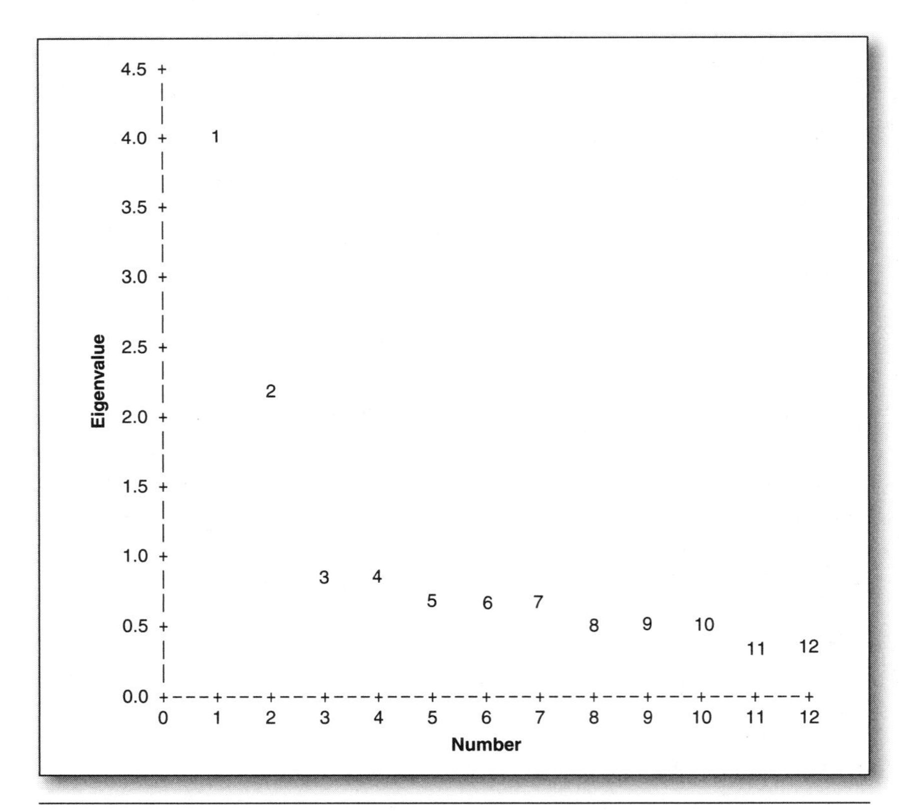

Figure 6.14 A scree plot from factor analysis of selected items

This strongly suggests that 2 factors account for much of the variation among the items.

Having determined how many factors to retain, we reran the program specifying two factors and Varimax (orthogonal) rotation. Had we failed to approximate simple structure, we might have performed an oblique rotation to improve the fit between items and factors. However, in this case, the simpler orthogonal rotation yielded meaningful item groupings and strong, unambiguous loadings. This is evident from the table of factor loadings, shown in Table 6.2. Each row contains the loadings of a given item on the two factors. An option available in SAS has reordered the items in the table so that those with high loadings on each factor are grouped together.

	Rotated Factor Pattern	
	Factor 1	*Factor 2*
Item I	.78612	−.22093
Item K	.74807	−.18546
Item D	.71880	−.02282
Item E	.65897	−.15802
Item G	.65814	.01909
Item A	.59749	−.14053
Item H	.51857	−.07419
Item F	−.09218	.82181
Item J	−.10873	.78587
Item C	−.07773	.75370
Item L	−.17298	.73783
Item B	−.11609	.63583

Table 6.2 Item Loadings on Two Factors

In this table, factor loadings greater than .50 have been underlined. Each factor is defined by the items that load most heavily on it (i.e., those underlined). By referring to the content of those items, one can discern the nature of the latent variable that each factor represents. In this case, all the items loading strongly on Factor 1 concern the parent as an influence over whether a child remains safe and healthy. Those loading primarily on Factor 2, on the other hand, concern the influence of luck or fate on the child's health.

These two homogeneous item sets can be examined further. For example, alpha could be computed for each grouping. Computing alpha on these item groupings using SAS yields the information shown in Table 6.3. Both scales have acceptable alpha reliability coefficients. Note that the SAS CORR procedure calculates alpha for unstandardized and standardized items. The latter calculation is equivalent to using the correlation-based alpha formula. For both scales, these two methods of computing alpha yield quite similar values. Note also that for neither scale would alpha increase by dropping any item. Alphas nearly as high as those obtained for the full scales result from dropping one item (i.e., Item H from Scale 1 and Item B from Scale 2). However, retaining these items provides a bit of additional insurance that the reliability will not drop below acceptable levels on a new sample and does not increase the scales' lengths substantially.

All the caveats applicable to scales in general at this point in their development are applicable to factor analytically derived scales. For example, it is very important to replicate the scales' reliability using an independent sample. In fact, it probably would be useful to replicate the whole factor analytic process on an independent sample to demonstrate that the results obtained were not a one-time chance occurrence.

SAMPLE SIZE

The likelihood of a factor structure replicating is at least partially a function of the sample size used in the original analysis. In general, the factor pattern that emerges from a large-sample factor analysis will be more stable than that emerging from a smaller sample. Inevitably, the question arises, "How large is large enough?" This is difficult to answer (e.g., MacCallum, Widaman, Zhang, & Hong, 1999). As with many other statistical procedures, both the relative (i.e., to the number of variables analyzed) and the absolute number of subjects should be considered, but factors such as item communalities also play a role (MacCallum et al., 1999). The larger the number of items to be factored and the larger the number of factors anticipated, the more subjects should be included in the analysis. It is tempting, based on this fact, to seek a standard

Cronbach's Coefficient Alpha for RAW variables: .796472; for STANDARDIZED variables: .802006				
	Raw Variables		*Std. Variables*	
Deleted Variable	*Correlation With Total*	*Alpha*	*Correlation With Total*	*Alpha*
ITEM I	.675583	.741489	.676138	.749666
ITEM K	.646645	.748916	.644648	.755695
ITEM E	.545751	.770329	.535924	.775939
ITEM D	.562833	.763252	.572530	.769222
ITEM G	.466433	.782509	.474390	.787007
ITEM H	.409650	.793925	.404512	.799245
ITEM A	.437088	.785718	.440404	.793003
for RAW variables: .811162; for STANDARDIZED variables: .811781				
	Raw Variables		*Std. Variables*	
Deleted Variable	*Correlation With Total*	*Alpha*	*Correlation With Total*	*Alpha*
ITEM F	.684085	.748385	.682663	.749534
ITEM C	.596210	.775578	.594180	.776819
ITEM J	.636829	.762590	.639360	.763036
ITEM L	.593667	.776669	.592234	.777405
ITEM B	.491460	.806544	.493448	.806449

Table 6.3 Coefficient Alphas for All Items and for All $k - 1$ Combinations of Items for Two Different Sets of Items

ratio of subjects to items. However, as the sample gets progressively larger, the ratio of subjects to items often can diminish. For a 20-item factor analysis, 100 subjects would probably be too few, but for a 90-item factor analysis, 400 might be adequate. Tinsley and Tinsley (1987) suggest a ratio of about 5 to 10 subjects per item up to about 300 subjects. They suggest that when the sample is as large as 300, the ratio can be relaxed. In the same paper, they cite another set of guidelines, attributed to Comrey (1973), that classifies a sample of 100 as poor, 200 as fair, 300 as good, 500 as very good, and 1,000 as excellent. Comrey (1988) stated that a sample size of 200 is adequate in most cases of ordinary factor analysis that involve no more than 40 items. Although the relationship of sample size to the validity of factor analytic solutions is more complex than these rules of thumb indicate, they will probably serve investigators well in most circumstances.

It is certainly not uncommon to see factor analyses used in scale development based on more modest samples (e.g., 150 subjects). However, the point is well taken that larger samples increase the generalizability of the conclusions reached by means of factor analysis. Of course, replicating a factor analytic solution on a separate sample may be the best means of demonstrating its generalizability.

CONCLUSION

Factor analysis is an essential tool in scale development. It allows the data analyst to determine the number of factors underlying a set of items so that procedures such as computing Cronbach's alpha can be performed correctly. In addition, it can provide us with insights into the nature of the latent variables underlying our items.

7

An Overview of Item Response Theory

Item response theory (IRT) is an alternative to classical measurement theory, also called classical test theory (CTT). IRT has received increasing attention in recent years and is often presented as a modern and superior alternative to CTT (e.g., De Boeck & Wilson, 2004; Embretson & Reise, 2010; Nering & Ostini, 2010). The two approaches share several fundamental characteristics. For example, scale unidimensionality is a requirement of both classical and IRT approaches. That is, items must share one and only one underlying variable if they are to be combined into a scale. Stated differently, under both approaches, the items must share a single common cause and thus be correlated with one another. If a set of items is multidimensional (as a factor analysis might reveal), then the separate, unidimensional item groupings must be dealt with individually. This is true under both classical and IRT approaches. So, for example, if a set of 50 items formed five separate factors, each of the five item clusters would be treated separately, potentially yielding five scales.

Both approaches to measurement also differentiate between item variation that arises from the actual levels of the latent variable and variation that arises from error, but the way that the distinction between true score and error is handled differs across the two measurement approaches. The basic idea underlying classical measurement theory is that an observed score is simply the result of the respondent's true score plus error. That error is not differentiated into subcategories, such as differences across time, settings, or items. Instead, all sources of error are collected in a single error term. IRT methods differentiate error more finely, particularly with respect to characteristics of individual items that may affect their performance.

A goal of IRT is to enable a researcher to establish certain characteristics of items that are independent of who is completing them. This is analogous to physical measurement in which one can assess an attribute of an object (e.g., length or weight) without regard to the specific nature of the object. Twenty pounds, for example, means the same thing no matter what is being weighed. Thus, a conventional bathroom scale yields information about a specific characteristic of objects (i.e., weight) irrespective of the nature of the object being weighed. IRT aspires to do the same thing with questionnaire items. Classical methods have an inherent linkage between the measurement tool and the people being measured that IRT approaches, at least in theory, do not. For example, a scale's reliability under classical measurement theory is influenced by the correlations among the items making up that scale. If the individuals whose data form the basis for that reliability assessment are extremely uniform in their levels of the attribute the scale measures, their range of true scores will be small. A consequence of that truncation of range is that the correlations among the items will be smaller and, hence, the scale reliability will be lower relative to a sample in which people vary more widely with respect to the measured attribute. Consequently, the reliability one obtains under classical measurement theory is not only about the performance of the measurement instrument but, under some circumstances, is also about the characteristics of the sample under study. IRT methods do not base reliability assessments on sample data in the same way that classical measurement approaches do. Of course, ultimately, for both approaches, item information is gathered from people. Consequently, whether the theoretical advantage that IRT has over classical measurement approaches is realized depends on the extent to which items under IRT are evaluated across large, heterogeneous samples.

Describing all the differences between classical and IRT approaches is beyond the scope of this chapter. For a relatively accessible overview of IRT, readers are referred to Hambleton, Swaminathan, and Rogers (1991). Rather than attempting an exhaustive comparison, in the sections that follow, I will focus on three key distinctions between classical and IRT-based measurement approaches: (a) the emphasis on items versus the scale as a whole, (b) the identification of items with specific levels of the attribute being measured, and (c) the visual representation of item and scale characteristics.

The first important distinction between the two approaches is that IRT pays considerable attention to the properties of individual items composing a scale. Classical approaches, in contrast, tend to emphasize characteristics of the scale as a whole. As an example of this difference, consider reliability. When we discussed Cronbach's coefficient alpha within the context of classical measurement, we observed that alpha can be enhanced either by increasing the number of items or by improving the average inter-item correlation. Thus, either more items or better items (in the sense of being more strongly associated with the latent variable) will improve reliability. In classical measurement theory, a scale's reliability is often increased by redundancy—adding more items. Typically, reliability is enhanced in IRT approaches not by redundancy but (where possible) by identifying better items. That is, IRT approaches view reliability as fundamentally about the individual items, whereas classical approaches tend to view it from the perspective of the scale as a whole. While classical methods, such as examining item-total correlations, can also determine how each individual item contributes to overall scale reliability, this process is less ingrained in classical than in IRT approaches to scale development and evaluation.

A second difference between the two measurement approaches is that IRT explicitly examines what level of the attribute being measured most strongly influences an item. Different items may be "tuned" to different levels of an attribute and, thus, may be sensitive to differences over different portions of the full continuum of that attribute. For example, an item such as "I sometimes feel sad" probably measures a lower level of sadness or depression than an item such as "I feel that life is not worth living." The first item may differentiate someone who rarely experiences feelings of sadness or depression from others who experience those feelings more frequently. It may not do a good job of differentiating people who occasionally are sad from those who very often are sad. On the other hand, the second item may differentiate only people at the upper extreme of the continuum. People along a substantial portion of that continuum may all fail to endorse this item, with only those who are substantially more sad or depressed answering affirmatively. IRT approaches would identify these two items as representing different points on a continuum of depression. By determining which items are relevant to which regions of the continuum for the variable being measured, IRT methods can help the test developer represent the full range of the continuum for the attribute being measured (depression in this example) with appropriate items. That is, items can be included in a scale that represents low, moderate, and high levels of the attribute more readily under IRT because the methods explicitly relate the individual items to the portion of the attribute continuum for which they are relevant. While there are classical measurement methods that can accomplish similar results (such as examining the proportions of respondents endorsing each item and the total scores that typify endorsers and nonendorsers), those methods are not a customary step in classical measurement methods. In contrast, assessing item "difficulty" is an integral part of most IRT approaches.

How strongly an item relates to the latent variable and where along the continuum an item falls both have an impact on reliability. Items strongly associated with the latent variable will also be strongly associated with one another, thus increasing the average inter-item correlation and scale reliability. Moreover, by developing good items along the full range of the variable, one can assure that the resultant scale is reliable *across a wide range of the attribute being measured* and not just at certain portions of the continuum. Thus, both of these common, item-oriented aspects of IRT can enhance the reliability of a scale.

A third contrasting feature of IRT relative to classical measurement methods is closely related to the first two. IRT approaches make extensive use of graphical depictions as a way of representing the properties of individual items and scales as a whole. We will look at some of these depictions shortly, but first, I will discuss the origins of some of the terms commonly used to discuss item properties under IRT and closely linked to the graphical depictions IRT employs.

Because IRT originated in the context of ability testing, its vocabulary contains terms usually associated with that content area. Also, because items on ability tests are usually graded as "correct" or "incorrect" (even though their original format may involve more than two response options), the classic applications and examples of IRT involve items that take on one of two states (e.g., "pass" vs. "fail"). It is easiest to discuss IRT initially for items of this type, although there is no reason why the methods stemming from the theory should not be extended (as, indeed, they have been) to items with other response formats (e.g., Likert scales) tapping other content domains. We will consider items of this latter type later in the chapter.

IRT is really a family of models rather than a theory specifying a single set of procedures. One important way in which the alternative IRT models differ is in the number of item parameters with which they are concerned. A common approach in recent years has been the three-parameter model, which, not surprisingly, concentrates on three aspects of an item's performance. These are the item's *difficulty,* its capacity to *discriminate,* and *guessing*—or more generally, its *susceptibility to false positives.* An early, and still popular, member of the IRT family is Rasch modeling (e.g., Rasch, 1960; Wright, 1999), which quantifies only the difficulty parameter.

ITEM DIFFICULTY

Although the term is a clear carryover from ability testing, the concepts it represents are more widely applicable. Item difficulty refers to the level of the attribute being measured that is associated with a transition from "failing" to "passing" that item. Most of us have seen old movies depicting carnivals or amusement parks featuring a certain feat of strength. The "measurement device" is a vertical track along which a weight travels. At the top of the track is a bell. Initially, the weight rests at the bottom of the track on the end of a plank serving as a kind of seesaw. "Respondents" strike the end of this seesaw opposite the weight with a large mallet, thus sending the weight flying upward along the track. Their goal is to propel the weight with enough force to strike the bell. For our purposes, we can think of the entire apparatus as the "item" (see Figure 7.1).

Figure 7.1 A hypothetical apparatus for testing strength in which striking the pad with the hammer with sufficient force causes the bell to ring

The *difficulty* of the item is the amount of strength the "respondent" must possess (or, more accurately, the force she or he must transmit) in order to "pass" the item (i.e., ring the bell). Clearly, one could construct different items with different degrees of difficulty (e.g., more difficult "items" with longer tracks or heavier weights). It should be possible, however, to calibrate the difficulty of a particular apparatus that is independent of any characteristic of the person who happens to be swinging the mallet at the moment.

Because this "item" is a physical object, it would be fairly easy to determine with reasonable precision how much force was needed to cause the bell to ring (ignoring, for the moment, the effect of striking the seesaw in slightly different locations relative to the fulcrum). So, the carnival operator could presumably order a 10-pound apparatus or a 100-pound apparatus to achieve either a high or low "passing rate" among people playing the game. Each might be specifically suitable for different groups of customers (e.g., children attending a school fair vs. adults attending an athletic camp).

One can characterize a questionnaire item in a similar way. Consider, for example, items measuring depression, as we did earlier. One could construct the item to be relatively "easy" (e.g., "I sometimes feel sad") or relatively "difficult" (e.g., "I feel that life is not worth living"). In the first instance, only a modest amount of the attribute "depression" would be needed to "pass" (i.e., endorse) the item. But would not the likelihood of the person having had that feeling depend on who was asked? If, for example, we posed the question to people who were clinically depressed, we would probably find a larger proportion of that sample endorsing the item than if we administered it to the general population. The goal of determining item difficulty is to establish in an absolute sense how much of the attribute is required to pass the item. If this can be done, then a person's passing the item has a constant meaning with respect to level of depression, irrespective of who that person is or the average level of depression in the sample under study. In other words, the person can be characterized not merely in reference to a specific sample but in terms of a metric independent of any specific sample.

ITEM DISCRIMINATION

The second parameter IRT addresses is the degree to which an item unambiguously classifies a response as a "pass" or "fail." Stated differently, the less the ambiguity about whether a person is truly a pass or fail, the higher the discrimination of the item in question. Using our carnival bell analogy, there may be occasions when the weight barely makes contact with the bell, causing observers to disagree as to whether the bell actually rang. Some might hear a faint ring, while others hear nothing. Within the range of force that propels the weight so that it touches the bell but does not produce what all agree is a clear ringing sound, the device is providing ambiguous information. Looking at this ambiguity another way, the same force applied many times might result in observers determining that the bell rang on some occasions but did not ring on others. A somewhat greater force will consistently produce an unambiguous ring, while a somewhat weaker force will produce an equally unambiguous failure to ring. But there is a small range of force in which the device is ambiguous. An alternative device might operate differently and produce less ambiguous results. For example, the weight touching the bell might close an electrical circuit that trips a relay, causing a light to illuminate and remain lit until it is reset. If well engineered, such a device would probably yield consistent results over a fairly small range of forces and, thus, would discriminate better than the standard device. Alternatively, a device that had no bell at all but instead required the observers to raise their hands if the weight crossed a predetermined line marked next to its track would probably produce more ambiguity and, thus, would discriminate less well. So, a device or item that discriminates well has a narrow portion of the range of the phenomenon of interest in which the results are ambiguous. A less discriminating device or item has a larger region of ambiguity.

GUESSING, OR FALSE POSITIVES

The third parameter in IRT is guessing, or false positives. The "guessing" nomenclature is, once again, a legacy of IRT's origins in ability testing. On a multiple-choice test, a respondent who did not actually know the correct answer could pass the item by guessing the correct response option. Such a guess is a false positive. That is, it yields a positive indication that the respondent possesses a certain level of ability even though the person does not actually know the correct answer to the item in question. Thus, regarding the parameter as reflecting a false positive rather than guessing allows us to generalize more easily beyond the ability-testing context. Another carnival analogy may clarify the idea of a false positive. You may have seen booths in which a person sits behind a protective barrier of some sort above a tank of water on a platform that is connected to a lever extending to one side on which a target is painted (see Figure 7.2).

Contestants are invited to throw baseballs at the target, which, if hit, causes the platform to collapse and the person to fall into the tank of water below him or her. We can think of this device as an "item" that measures throwing accuracy. Causing the person on the platform to fall into the tank of water constitutes a "pass" on this item. (By now, you should be able to describe how

Figure 7.2 A hypothetical apparatus for measuring throwing accuracy in which striking the target with a ball causes the platform to collapse and the person sitting on it to fall into a tank of water

variations in the apparatus could increase or decrease the *difficulty* and *discrimination* of the device.) With this particular device, one can imagine how "false positives" might occur—that is, how a respondent with virtually no ability could score a "pass" by causing the person perched above the tank to be dunked. One way might be that the "respondent" throws wildly but the ball just happens to hit the target (it has to go somewhere, after all). Or, alternatively, the apparatus might malfunction and the platform might collapse spontaneously. In these cases, the player/respondent would "pass" based not on ability but on some unrelated circumstance. Thus, it is possible to "pass" this test of throwing accuracy even if one has little or no ability. As noted earlier, in the context of ability testing, false positives most commonly occur as the result of successfully guessing the correct response to a question despite not really knowing the answer. (In measurement contexts where the opportunities for guessing or other types of false positives are minimal, such as using scales to measure weight, a two-parameter model is often sufficient.)

Each of these three item parameters—difficulty, discrimination, and false positives—bears a fairly obvious relationship to measurement error. If (a) the difficulty of an item is inappropriate, (b) the area of ambiguity between a pass and a fail is large, or (c) the item indicates the presence of a characteristic even when it is absent, then the item is prone to error. IRT quantifies these three aspects of an item's performance and, thus, provides a means for selecting items that will likely perform well in a given context. Now, let's see how these three parameters relate to the third important feature that differentiates IRT approaches from classical methods: the use of graphical depictions to illustrate item properties.

ITEM-CHARACTERISTIC CURVES

An item's difficulty, discrimination, and generation of false positives can be summarized in the form of an item-characteristic curve (ICC) that graphically represents the item's performance. Typically, an ICC is roughly S-shaped, and different parts of the curve reveal information about each of the three parameters of interest.

Figure 7.3 shows what an ICC might look like. The X axis represents the strength of the characteristic or attribute being measured (e.g., knowledge, strength, accuracy, depression, social desirability, or virtually any other measurable phenomenon). The Y axis represents the probability of "passing" the item in question based on the proportions of failing and passing scores observed. Seeing how an ICC can be used to assess item quality is actually easier if we look at a diagram representing two items that we can compare.

Figure 7.4 illustrates item difficulty by showing two curves. Note that the points at which the curves attain a 50% probability of their respective items being passed are different. For the lighter curve, that point is farther to the right. That is, the amount of the attribute must be higher for an individual to have a 50% chance of passing the item represented by the lighter line than the item represented by the darker line. Using that criterion, the item represented by the lighter line is more difficult. Difficulty in this case is not a subjective judgment but a factual description of the point on the X axis corresponding with the curve's crossing of the .50 probability value on the Y axis.

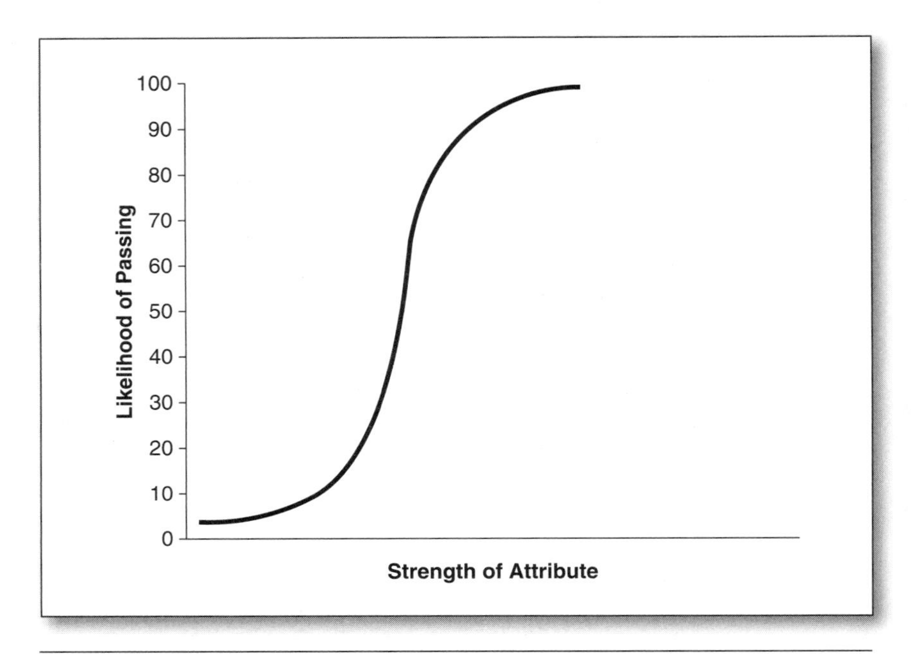

Figure 7.3 Hypothetical ICC for a pass-fail item

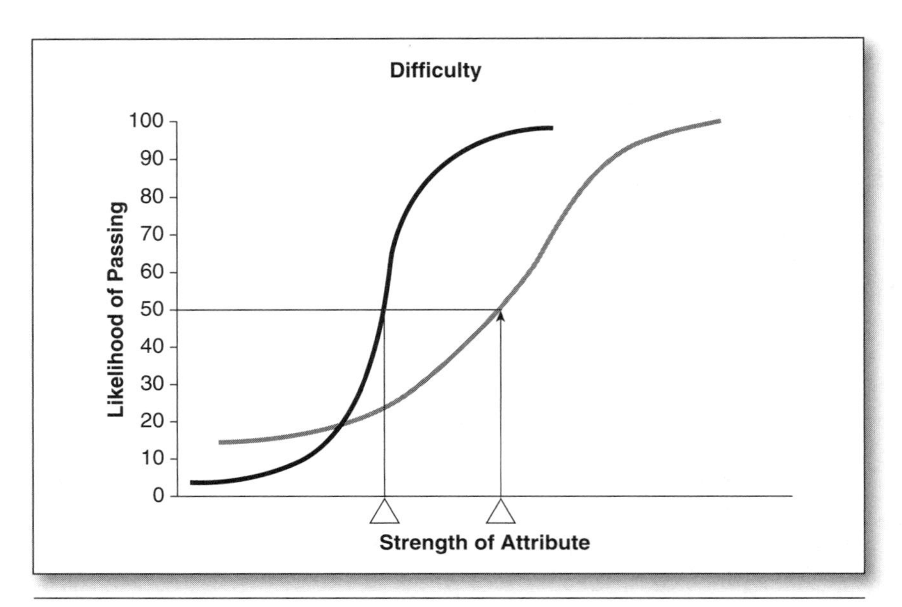

Figure 7.4 Two hypothetical ICCs for dichotomous items differing in difficulty

Figure 7.5 illustrates how we assess discrimination using the same two ICCs. The item corresponding with the darker curve has a steeper slope at the 50%-pass point than does the item represented by the lighter curve. A consequence of this is that a smaller increase in the attribute yields a greater increase in confidence that a respondent will pass this item than is the case for the item represented by the lighter line. So, the steeper dark line reveals that the region of the X axis that corresponds with an ambiguous score is smaller than the equivalent region for the other item. Thus, the dark-line item discriminates more effectively than the light-line item between those who fail and those who pass.

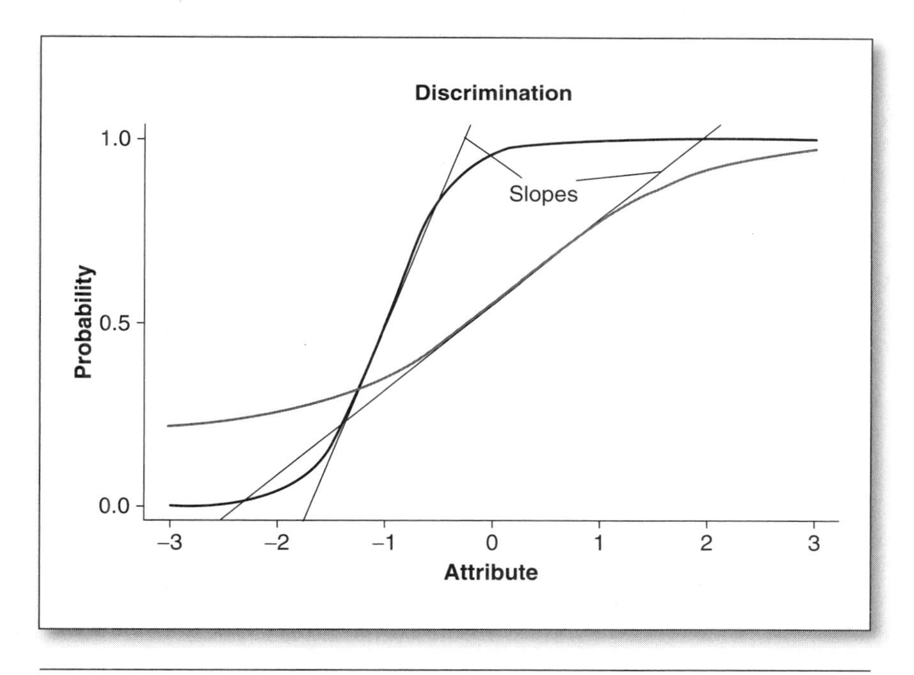

Figure 7.5 Two hypothetical ICCs for dichotomous items differing in discrimination

Finally, in Figure 7.6, we can look at the two items' propensities to yield passing grades even when the ability (or whatever attribute is being measured) of the respondent is essentially zero. As you might have guessed, this is determined by the point at which the ICC intersects with the Y axis. For the dark-line item, the intercept value is zero. Thus, the probability of a person passing the item if he or she completely lacks the attribute in question is quite small.

For the light-line item, there is a substantial probability (about 20%) that someone with no ability will pass the item and, thus, be indistinguishable, based on the item in question, from someone with high ability. (Note that this would be the case with a multiple-choice item having five possible answers, one of which was correct.) The corresponding diagram points out the differences in the Y intercepts of the two items—the basis for concluding that, once again, the dark-line item is the better performer.

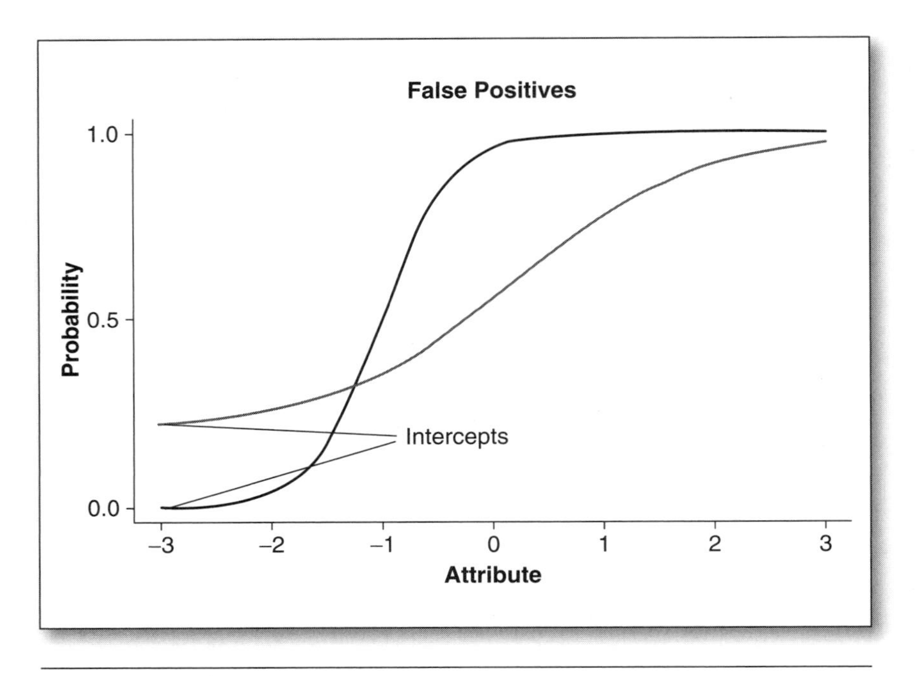

Figure 7.6 Two hypothetical ICCs for dichotomous items differing in false positives

In theory, one can use IRT to establish the parameters for each of many items. Then, depending on the details of the intended application, optimally performing items can be used to address the measurement problem at hand. For example, "easy" items could be assembled for use with respondents who possess relatively low levels of the ability in question and difficult items for those possessing high levels. This is directly analogous to using the 10-pound bell apparatus at a fair or carnival intended for children and the 100-pound apparatus at a camp for adult athletes. Using the wrong item—like using the wrong bell apparatus—can lead either to frustration (if the task is too difficult) or a lack of motivation (if the task is too easy). Also, if the measure being compiled will be the basis of important decisions, then minimizing the range of ambiguity for each item and the likelihood of false positives are also attractive possibilities.

An extension of this idea is the use of adaptive testing. In adaptive testing, the items administered to each individual can be tailored to that person's level of the attribute. Whether or not a previous item was passed can guide the selection of the next item, giving preference to an item that is tuned to the ability level that prior responses suggest the respondent has. This process is typically managed by a computer and is often referred to as computerized adaptive testing, or CAT. Although this methodology requires a large pool of items representing a wide range of ability levels, it is a powerful tool that IRT approaches make possible.

IRT approaches have the distinct advantage of directing our attention to three (in the three-parameter version that currently is popular) important aspects of an item's performance. With methods rooted in classical measurement theory, we may know (e.g., from its performance in factor analysis or coefficient alpha computations) if an item is performing well or poorly but may not have as clear an understanding of the nature of any deficiencies it possesses. IRT, in contrast, may help us assess an item's strengths and weaknesses more specifically.

IRT APPLIED TO MULTIRESPONSE ITEMS

The introduction to IRT presented above applies to three-parameter models and deals with dichotomous responses, such as True-False. As I suggested earlier, there are several IRT models. Commonly in the social and behavioral sciences, we deal with ordered, multilevel responses of the types we have typically discussed in previous chapters.

For instruments with multilevel response options, such as Likert scales, specially adapted IRT models apply. One of the most common of these is the *graded response model* developed by Fumiko Samejima (e.g., Samejima, 1969) and commonly applied by means of David Thissen's Multilog software program (Thissen, Chen, & Bock, 2003). This model provides information about how each of the multiple-response options relates to ability. For a good item, each response option should occupy a more-or-less distinct portion of the ability continuum.

Imagine, for example, an item such as "I experience dizziness when I first wake up in the morning" with response options of (0) "never," (1) "rarely," (2) "some of the time," (3) "most of the time," and (4) "almost always." One would expect that the likelihood of endorsing progressively higher options would be associated with progressively higher levels of morning dizziness (i.e., progressively higher levels of the attribute being assessed). In a way, an item such as this takes the place of several binary items. So, instead of separate yes-no items such as "I *never* experience dizziness when I first wake up in the morning," "I *rarely* experience dizziness when I first wake up in the morning," "*Some of the time* I experience dizziness when I wake up in the morning," and so forth, one item is presented with several response options from which the respondent can choose.

If the item worked well on a given population, the likelihood of choosing the first response option (i.e., "never") should be high for people with low levels of the attribute (dizziness) and should diminish virtually to zero as the level of the attribute increases. So, the curve representing the likelihood of selecting this first response option should be high at its leftmost extreme (the region of the continuum corresponding with low levels of morning dizziness) and low at its rightmost extreme. At the other end of the scale, the likelihood of choosing the last response option (i.e., "almost always") should be virtually zero for people at low-to-moderate levels of the attribute, progressing as dizziness increased and approaching 100% for people with the highest levels of morning dizziness. The curve for this response should be very low at lower ends of the continuum of morning dizziness, rising to a maximum value at the rightmost extreme of the continuum. Thus, the curves for both of these extreme response options are nonsymmetrical, with one end very low and the other end very high. Also, each of these two curves should be positioned at an extreme of the continuum corresponding with the level of the attribute the response option represents. The curves for the in-between response options, by contrast, ideally should be more or less symmetrical, with low likelihood of endorsement by respondents at the extremes of the attribute continuum and peak likelihood at some point on that continuum that is appropriate for the response option in question. For higher response options, the peak should occur at higher levels of the attribute continuum than for lower response options.

We can depict the likelihood of choosing each of the response options, based on one's level of the attribute being measured, with a series of *category-response curves.* A complete set of hypothetical category-response curves for five response options used in the earlier example might look something like the one shown in Figure 7.7.

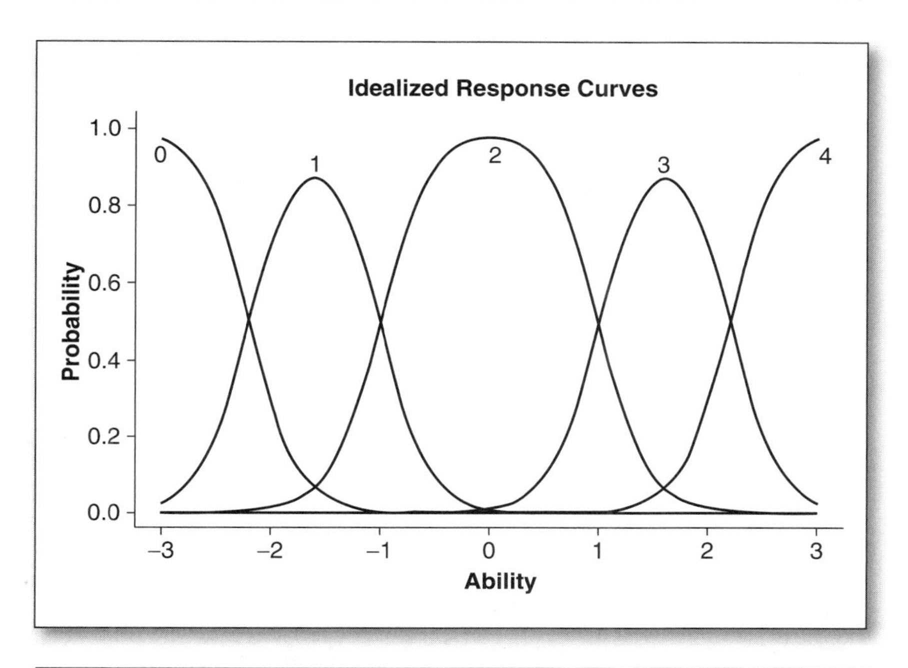

Figure 7.7 Idealized category-response curves for five response options

In the idealized illustration shown in Figure 7.7, each labeled curve corresponds with a response option in the morning dizziness example given previously. The baseline represents the strength of the attribute being measured by the item. The ability scale is centered on 0 and can be interpreted in a way analogous to standardized scores. Note that this illustration is an idealized example and not an actual graph from real data. This example retains a characteristic of true item-response curves: For any level of the attribute, the sums of the various curves at that point on the baseline equals 1.0. So, for a vertical line drawn at any point on the attribute scale, the values (from the left-hand axis) of the points at which it crosses the response category curves will sum to 1.0. In other words, the probability of choosing *some* response equals unity for any respondent with any level of the attribute being assessed.

A less idealized but still hypothetical example is illustrated in Figure 7.8. The location of each response option's curve along the attribute dimension indicates the "difficulty" of that response option. We can see that the peaks for each successive response option curve are ordered as we would expect, with the response options indicating less morning dizziness to the left of those indicating more. That is, indicating more frequent dizziness is associated with a curve positioned higher (farther to the right) along the attribute continuum, as we would hope.

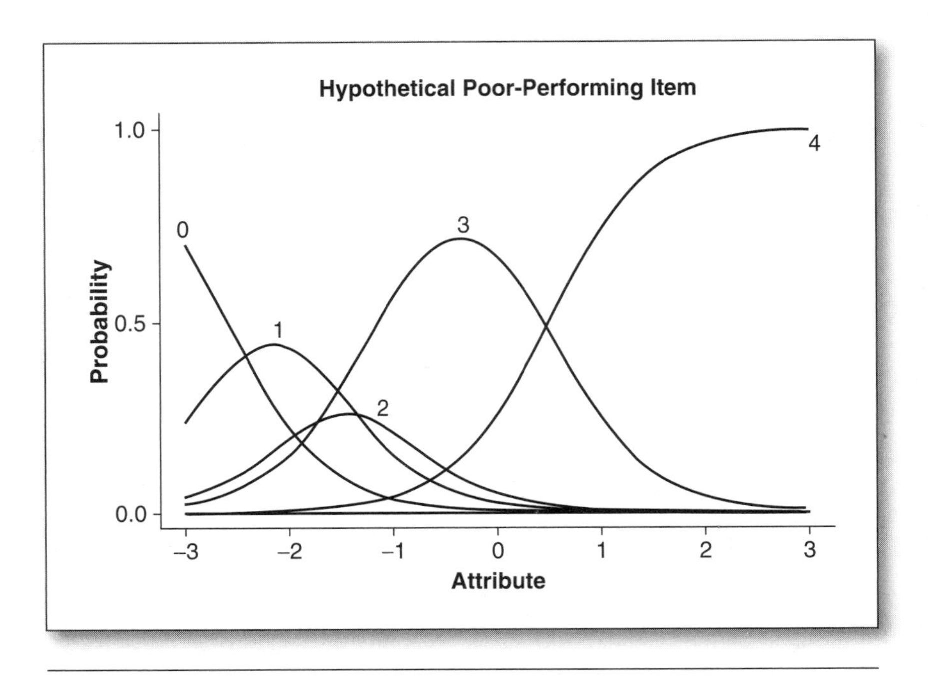

Figure 7.8 Hypothetical example with category-response curves crowded at the lower end of the attribute scale

Another notable feature of Figure 7.8 is that response curves seem to be crowded toward the left side of the scale. This means that the question differentiates more finely among people with relatively little dizziness than it does among people with relatively much dizziness. If we consider the range on the attribute scale between –1.5 and –.5, we see that all five of the response options are occasionally selected by respondents with that range of dizziness values, although response Options 1 and 3 are most often chosen. Moreover, within that 1-point range, those at the higher end are most likely to choose Option 3 ("most of the time") whereas those at the low end will more often choose Option 1 ("rarely"). Thus, even within that narrow range, we can still make differentiations between people who are higher and lower with respect to dizziness.

In contrast, if we look farther up the attribute scale, at the range between 2.0 and 3.0, the overwhelming likelihood is that respondents with dizziness levels anywhere between those values will choose Option 4 ("almost always"). In fact, "almost always" is the most likely response for any respondent scoring above about .8 on the attribute scale. Thus, this item provides little differentiation among people varying between .8 and 3.0 with respect to their levels of dizziness.

The item-response curves in Figure 7.8 have even more to tell. Note that response Option 2 ("some of the time") overlaps quite a bit with 1 ("rarely") and 3 ("most of the time"). There is virtually no portion of the ability range in which the response "some of the time" is the most likely to be chosen. In essence, the "some of the time" response option is not doing anything that the response options to either side of it cannot accomplish. A respondent who possessed a degree of morning dizziness comparable to a score of –1.5 on the attribute scale (the point where respondents are most likely to choose response Option 2) would still be more likely to choose either response Option 1 or 3. Choosing Option 2 in this hypothetical case amounts to equivocating between rarely having fatigue and sometimes having fatigue. Considering this item in isolation, without regard to any other items that might be included with it in a scale, dropping the "some of the time" response option would probably make sense.

The shape of the individual curves also gives us information about discrimination. This is most obvious for the last curve, which most closely resembles the curves we examined for the three-parameter, binary response model (as shown in Figure 7.3). The leftmost curve is sort of a crude mirror image of what we saw under the three-parameter model in that it attains its highest level at the extreme left and descends as it progresses up the attribute scale, but again, the steepness of its slope is indicative of its discrimination. For the curves in between, steepness (or, perhaps more accurately, kurtosis) indicates discrimination. Response options whose curves are tall and pointy discriminate better than ones whose curves are more broad and flat. In Figure 7.8, response Option 2 ("some of the time") has a curve that is fairly squat, indicating that the response option does not discriminate well. This further supports the notion that eliminating that option might be appropriate.

To conclude this section, we will look at response-category curves generated from actual data (provided by and used with the permission of my UNC colleagues Darren DeWalt and David Thissen) and will use those curves to illustrate another purpose they can serve. Figure 7.9 represents an item administered to a group of female pediatric patients as part of an effort to develop an item battery to assess depression. The item asked the children to indicate how frequently they had cried more than usual in the recent past. Similar but not identical to our previous examples, the response options were 0 ("never"), 1 ("almost never"), 2 ("sometimes"), 3 ("often"), and 4 ("almost always"). The clustering of curves to the right suggests that respondents had to be fairly high on the attribute (i.e., depression) in order to provide an answer other than "never."

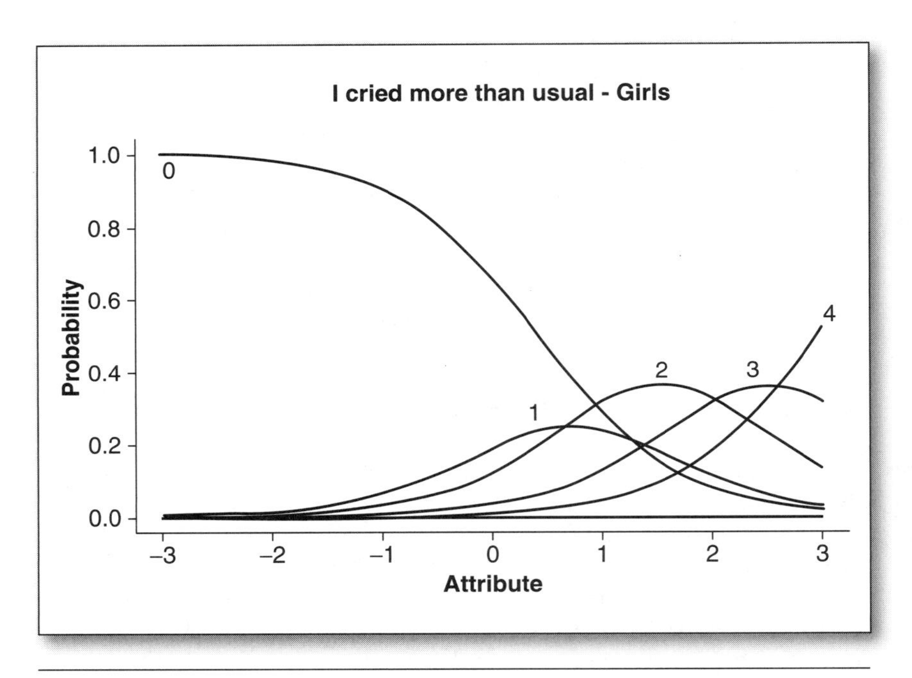

Figure 7.9 Category-response curves for girls on an item concerning crying more than usual

Figure 7.10 superimposes the responses of boys over those of girls. While this figure is quite busy, it reveals that the curves for boys and girls are generally similar but by no means identical. To make one critical difference between boys' and girls' responses clearer, in Figure 7.11, I have edited out the curves representing the three middle response options, leaving only those for "never" and "almost always." What this edited figure reveals is that, for Option 0, the boys' curve is higher than the girls', while for Option 4, the girls' is higher than the boys'.

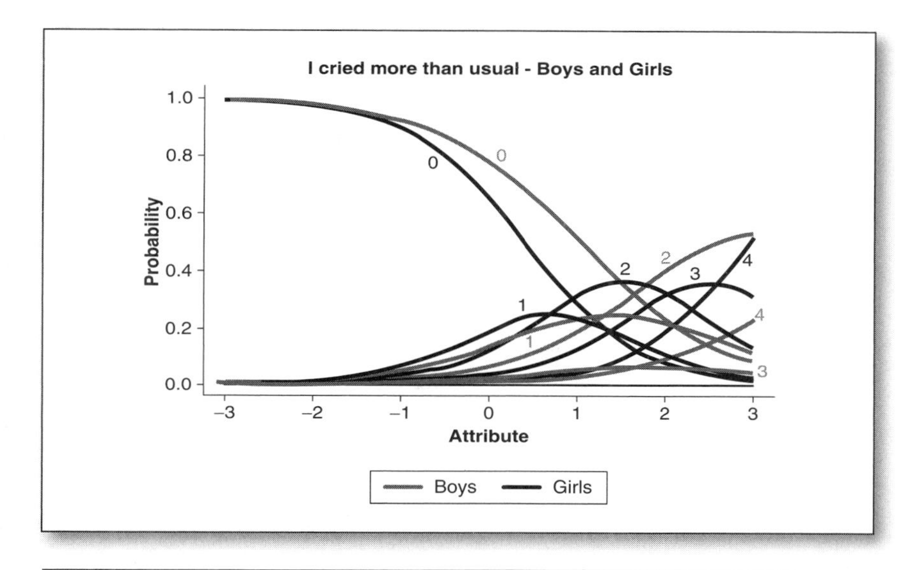

Figure 7.10 Superimposed category-response curves for boys and girls on item concerning crying more than usual

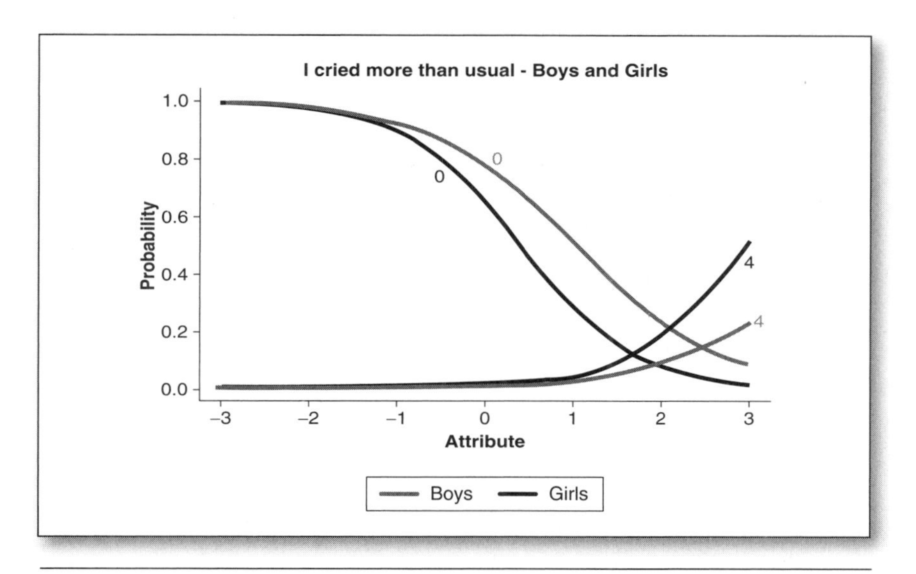

Figure 7.11 Superimposed category-response curves for girls and boys showing only the extreme response options

How should we interpret this difference? It suggests that across essentially the full range of values for the attribute (depression), boys are more likely than girls to say that they *have not* (i.e., to choose the "never" response option) cried more than usual. At the upper end of the attribute scale, girls are more likely than boys to acknowledge that they cried more than usual "almost always" over the time interval in question. This pattern suggests one of two things: Either (a) girls are more depressed than boys or (b) a lower level of depression generates higher responses for girls than it does for boys. That is, either the item is reflecting true gender differences between boys and girls or the item does not perform equivalently for the two genders. Our suspicion was that admitting to crying may be more difficult for boys than girls. As a consequence, boys who choose any response option other than "never" might actually be more depressed than girls choosing the same response option. This suggests that the item is manifesting differential item functioning (DIF). That is, different groups of respondents (boys and girls in this case) with identical levels of the attribute obtain different item scores, indicating that the item maps onto the attribute scale differently as a function of gender group. By looking at other items measuring depression that did not ask about crying and by observing essentially similar scores for boys and girls, we found further evidence of DIF for the item in question. While there are other methods for identifying DIF, many investigators find the visual information that IRT packages produce to be especially compelling.

In summary, inspecting response-category curves can provide information of several types about how an item is performing. Among the many things an inspection of response-category curves can reveal are (a) information about how fully an item represents the full range of the attribute being measured, (b) if the ordered response options accurately represent progressively higher regions of the attribute scale, (c) whether all response options are necessary and useful, (d) how well the item discriminates at various levels of the attribute, and (e) if an item manifests DIF. Other types of graphical output can be obtained from many IRT packages, and the visual representation of item performance is among the most useful features of IRT.

Complexities of IRT

While very attractive, IRT is not a quick solution to measurement problems. Like classical measurement theory, IRT does not determine the characteristics of items; it merely quantifies them. So, the technology per se allows a researcher to assess item performance but does not directly cause one to write better items or make poorly constructed items suddenly work well. Also, the assessment process can be daunting when using methods based on IRT. Classical measurement trades precision for simplicity by adopting a less differentiated but more tractable conceptualization of error sources. IRT makes the opposite choice, gaining precision but sacrificing simplicity. The application of IRT methods, furthermore, requires a considerable degree of expert judgment. These methods are still in an active stage of development, with new issues popping up and new solutions being offered.

To have confidence that the characteristics of items are being assessed independently of the characteristics of the sample being studied, a primary objective of IRT, one must demonstrate that those characteristics are consistent across a wide variety of respondents differing in various respects, including ability level. It is important that the item characteristics not be associated with other attribute-independent sample characteristics, such as gender, age, or other variables that should be uncorrelated with the one being measured. Item scores should vary only when the attribute of interest varies, not because of differences in any extraneous variables. So, for example, if we assume that spelling ability is unrelated to gender, we would have to demonstrate that boys and girls with equal ability would have the same probability of passing an item. If this were not true, then gender or some factor other than spelling ability would be influencing the item. Also, as with classical theory, items being examined in a set (i.e., making up a scale for measuring a variable) must share only a *single* underlying variable.

One consequence of these requirements is that researchers must have access to large numbers of diverse respondents in order to characterize items accurately. A central goal of IRT is to relate items to specific levels of the attribute being assessed. Good items and large samples are necessary for this to be achieved. The theoretical independence of item characteristics and sample characteristics that is a hallmark of IRT requires that items have been evaluated across the full range of the attribute of interest and that the measurement tool includes items that collectively are sensitive to all levels of that attribute.

How does an instrument developer ascertain the true level of the attribute in a way that allows for the generation of the ICCs? Returning to our hammer-and-bell analogy, how do you define strength in order to determine how much strength it takes to ring the bell on a particular piece of equipment? In many cases, if the true level of the attribute were knowable in some manageable fashion, there would be little need to develop a new measure. In theory, given responses from a large number of people to a fixed set of items, a computer program should be able to sort out differences arising from item versus person characteristics. To return again to the analogy of the carnival devices (bell-ringing apparatus and dunking machine), if enough people use two of each device, for example, it should be possible to determine which version of each type of device is harder and, also, to judge the skills of individuals at these two tasks. In practice, there is often a back-and-forth, iterative process involving administering items to gauge the level of the attribute for particular respondents, and then using that attribute estimate as a guide in determining the characteristics of other items. When the best items are identified on that basis, they can be used to obtain an improved estimate of individuals' levels of the attribute for the next round of item selection, and so on.

Given the nature of these processes, it is easy to see why IRT has been most enthusiastically embraced by commercial organizations concerned with ability testing, such as those administering the Graduate Record Examinations. The constant administration and evaluation of items over time provides an excellent basis for finding items whose characteristics are stable across variation in a wide range of other respondent characteristics.

Another complexity of IRT concerns the use of different items to measure the same thing at different times. Earlier, I briefly mentioned computerized adaptive testing (CAT; Van der Linden & Glas, 2000) as an approach to data collection that is often associated with IRT. In CAT, items are individually calibrated to the attribute scale using IRT methods and, thus, they can be selected for administration so that they match the attribute level of a particular respondent. This implies that different items will be optimally appropriate for different attribute strengths. Such differences can occur either between people who differ with respect to the attribute or within a given individual whose level of the attribute changes.

Consequently, using CAT often entails scores from different items being compared with one another. For example, if an attribute changes as a result of an intervention, the best items for measuring that attribute in any individual may be different after the intervention than before. For the consumer of a scientific report that bases a claim of intervention success on a comparison between different items across (because of initial differences in the attribute) and within (because of changes resulting from the intervention) individuals, trusting the data may require a leap of faith. Most of us have far more experience (and, perhaps, comfort) with circumstances in which the same items were given before and after an intervention than with the scenario I have just described. When IRT methods have been properly applied and items have been carefully calibrated to the attribute continuum on the basis of data collected from large, diverse samples, scores from different items can be converted to a common attribute metric in the same way that measurements taken in inches or millimeters can be converted to a common metric for length. When the precision with which items can be mapped onto the attribute dimension is low, however, interpreting scores based on different item sets may be problematic. Furthermore, because consumers of research reports may not fully understand all the issues involved in the type of item-dissimilar comparisons that CAT may involve, the authors of those reports may face a greater degree of skepticism.

CONCLUSIONS

Measurement methods based on IRT have many attractive features. But developing good items is hard work, irrespective of the theoretical framework that guides the process. Writing items that consistently measure the attribute under study and are insensitive to other respondent characteristics is not a trivial task. Whereas, in a measure based on classical theory, having multiple items can offset the imperfections of some of those items to a degree, the logic of IRT has each individual item standing and being judged on its own. Because one *can* discover, by examining ICCs, for example, that an item performs well does not mean that one *will.* Having credible independent knowledge of the attributes being measured is a requirement of IRT that is difficult to satisfy strictly but that can be adequately approximated with repeated testing of large and heterogeneous samples. When this is not an option, it may be exceedingly difficult to convince critics that the assumption has been adequately met.

My personal view is that where the assumptions of classical measurement theory are applicable (i.e., where the items are intended as equivalent indicators of a common underlying variable), the tractability and performance of such measures make them attractive options. On the other hand, if large samples are available, CAT is a preferred mode of administration, or score equivalence across multiple studies is of paramount importance, then the added complexity of IRT-based approaches may be the best choice. The mere use of those methods, however, is by no means a guarantee of the desired end product. The researcher must demonstrate that the assumptions of the chosen method have been met within acceptable limits and that the resultant measurement tool's reliability and validity are empirically verifiable.

Does IRT render classical methods obsolete? Many advocates of IRT recognize that both classical measurement theory and IRT have a role to play. For example, Embretson and Hershberger (1999), in the first of their recommendations for changes to current measurement approaches, state that "IRT and CTT approaches should be integrated in a comprehensive approach to measurement issues" (p. 252).

At least two large empirical studies (Fan, 1998; Stage, 2003) that compared characteristics of educational tests based on classical methods and IRT concluded that the classical methods performed as well as or better than IRT did. Fan's (1998) study was based on 40 samples of 1,000 individuals each, drawn from a pool of roughly 193,000 Texas schoolchildren tested in reading and mathematics skills. Item and person characteristics showed substantial consistency across methods. Stage (2003) compared classical and IRT methods using data from the Swedish Standard Aptitude Tests. Based on a sample of 2,461 test takers randomly drawn from a pool of 82,506, she concluded that while a three-parameter IRT model fit the data poorly, a model based on classical test theory performed quite well. More recently, Silvestro-Tipay (2009) conducted a similar but smaller study using test scores from 326 college freshmen. He concluded,

> The findings here simply show that the two measurement frameworks produced very similar item and person statistics both in terms of the comparability of item and person statistics, difficulty level of items, internal consistency, and differential item functioning between the two frameworks. (p. 29)

In a similar vein, two researchers from Educational Testing Service and the College Board recently presented a paper describing the results of a simulation study comparing IRT and CTT approaches with specifying test forms (Davey & Hendrickson, 2010). They concluded that "both approaches offer an interesting mix of both theoretical and practical advantages and drawbacks, with neither emerging as a clear favorite on paper." They further note that

> the most notable result of the study was the relatively small difference evident between the performance of the classical and IRT assembly methods. . . . The similarity in performance across methods may mean that the practitioner's choice depends largely on preference or operational convenience. (p. 2)

These reports should not be interpreted as suggesting that IRT has no advantages in any situation. None of these studies is definitive, and some differences were identified when the methods were compared. IRT analyses in some cases provided more detailed information than classically based analyses. As we saw in the graphs for multiresponse items, IRT methods can yield useful insights, such as indicating that not all response options are necessary or that an item does not capture information across the full range of the attribute being measured. IRT approaches also afford clear theoretical advantages (such as the independence of item characteristics from sample characteristics), and as methods are improved, some of its practical barriers may be lowered. But at a minimum, these studies argue that IRT is not *necessarily* superior to classical measurement methods. In short, reports of the demise of classical measurement are premature.

Addressing the virtues and vices of the two approaches, Zickar and Broadfoot (2008) note,

> Just as CTT has its own limitations, researchers have noted severe limitations of IRT that make the use of its methods difficult, impossible, or impractical in certain scenarios. These limitations include the need for large sample sizes, strong assumptions of unidimensionality [which is also true under classical methods], and difficulty running programs. (p. 48)

Later in the same volume, these authors state,

> Although the urban legend is that CTT is dead, we believe that there are many scenarios in which it would be preferable to use. Most of the reasons can be categorized due to limitations in data that might preclude IRT and practical considerations that might make CTT more preferable. (p. 50)

IRT will continue to increase in popularity. It will have distinct advantages over earlier methods in certain circumstances. IRT and classical methods will coexist, much as regression analysis shares the stage with structural equation modeling approaches. Although both IRT and structural equation modeling have carried things further than their antecedents, the earlier methods retain their utility.

8

Measurement in the Broader Research Context

The opening chapter of this volume set the stage for what was to follow by providing some examples of when and why measurement issues arise, discussing the role of theory in measurement, and emphasizing the false economy of skimping on measurement procedures. In essence, it sketched the larger context of research before the focus shifted to the specific issues covered in later chapters. This chapter returns to the big picture and looks briefly at the scale within the larger context of an investigation.

BEFORE SCALE DEVELOPMENT

Look for Existing Tools

Early in this volume, I suggested that scale development often arises from a lack of appropriate existing instruments. It is important and efficient to be quite certain that a suitable measurement alternative does not already exist. Elsewhere (DeVellis, 1996), I have suggested ways of searching for appropriate scales. Essentially, this process involves searching published and electronic compendia of measures to determine whether a suitable instrument already exists. Published series such as the *Mental Measurements Yearbook* (e.g., Spies, Carlson, & Geisinger, 2010) and *Tests in Print* (Murphy, Spies, & Plake, 2006) contain primarily clinical measures, including tests of ability and personality. These are often the tools that applied psychologists will use for assessing clients. Tools intended primarily for research are less prominent, but some are included. Another category of resources is targeted compilations, such as the classic *Measures of Personality and Social Psychological Attitudes* (Robinson, Shaver, & Wrightsman, 1991). Relevant journals are also an excellent place to find what measurement strategies have worked successfully for others interested in the same constructs.

With increasing frequency, researchers can find information about measurement instruments on the Internet. In fact, the Internet may be where the most rapid expansion of measurement-related information is appearing. As an example of the expansion of Internet resources, both the *Mental Measurements Yearbook* (Spies et al., 2010) and *Tests in Print* (Murphy et al., 2006) are now searchable online at the website for the Buros Institute of Mental Measurements at the University of Nebraska at Lincoln (http://www.unl.edu/buros/). Another notable example of an online resource for information about measures is the website associated with the Patient-Reported Outcomes Measurement Information System (PROMIS). PROMIS is a research initiative supported by the National Institutes of Health as part of its Roadmap for Medical Research (see Ader, 2007, for an overview) to bring greater order and clarity to the assessment of important end states in health research (PROMIS, 2010). This site is a gateway to information developed through a collaborative effort involving multiple universities and sponsored by the National Institutes of Health. The PROMIS investigators have reviewed, evaluated, classified, modified, and developed items assessing health outcomes in five broad domains: physical functioning, social functioning, emotional distress, pain, and fatigue. These items have undergone careful scrutiny using methods based in item response theory. The items in the various domain banks may represent the most carefully studied items of their type. As the PROMIS effort continues, items will undergo extensive validation and new item sets will likely be added.

Although there are exceptions, such as those noted above, Internet-based repositories of widely used measurement scales tend to be transitory. Exigencies of funding, staffing, and other administrative issues complicate the maintenance of such sites, and, as a result, many do not last more than a few years. Nonetheless, the Internet remains a valuable resource for researchers seeking to locate scales. Internet search engines are an excellent tool for identifying measurement instruments targeting specific variables. In a sense, advances in search algorithms have reduced the necessity for localized repositories. If a tool has been developed to measure a particular variable, an Internet search is likely to find either the tool itself or information about how it can be located.

As with any Internet-based information, the consumer needs to consider the origin of the information and its trustworthiness. Sites sponsored by universities, government agencies, or other established institutions or organizations usually offer accurate and credible information. In general, however, caution is warranted. Just as there is an abundance of "junk science" books, there are also plenty of Internet sites that may adopt a tone and appearance of scientific legitimacy that is not commensurate with their content. The skills you have gained from this book should help you evaluate measurement information sources more critically in any format and determine whether the measures described have demonstrated adequate reliability and validity. Whether or not the scale has been published in a peer-reviewed journal is yet another means of evaluating its credibility. Of course, in addition to having passed peer review, a scale will be useful only if it fits the researcher's conceptualization of the variable of interest. Thus, scrutiny beyond the imprimatur of appearance in a journal is warranted before deciding that a particular scale is a good choice for a given research project.

View the Construct in the Context of the Population of Interest

It is often important to assess whether the theoretical constructs we as researchers identify correspond with the actual perceptions and experiences of the people we plan to study. Focus groups (e.g., Krueger & Casey, 2000) and other qualitative approaches (e.g., DeWalt, Rothrock, Yount, & Stone, 2007) can be a means of determining whether ideas that underlie constructs of interest make sense to respondents. As an example, consider attributions—the explanations or interpretations people ascribe to various outcomes. Often, these are assessed along dimensions such as "under my control" versus "not under my control," "specific to this situation" versus "applicable to most situations," or "a characteristic of me" versus "a characteristic of the environment or situation." Research into attribution processes has been highly productive. Most people can analyze outcomes, such as receiving a job offer following an interview, along such dimensions. In some situations, however, this may not be the case. For example, asking non-native-born, rural-dwelling, poorly educated elders, unaccustomed to thinking in such terms, to evaluate illness outcomes or purchasing decisions along these three dimensions may not work well. Experience suggests that they may simply not understand the task because it is so foreign to the way they think about things. A focus group asking potential research participants to discuss relevant concepts might make such a problem evident and preclude a measurement strategy that would be doomed to failure.

Focus groups can also reveal the natural, everyday language that people use to talk about a concept. A young mother may not use the same terms as a marketing expert to describe reactions to a product. The former may use "pretending" to describe her child playing without using a particular toy, while a marketing researcher might describe such play as "product nondirected." Structuring items around the mother's language usage (e.g., "How much playtime does your child spend just pretending, without using any toys?") rather than the expert's (e.g., "How much time does your child spend in product-nondirected play?") is more likely to yield a tool suitable for measuring her perceptions of how her child interacts with various products.

Cognitive interviewing is another qualitative method for determining how potential respondents interpret and understand items. Although there are a variety of approaches (e.g., Willis, 2005), the basic idea is to learn from respondents by asking them what they understood an item to be about and how they formulated a response to it. This method can often reveal confusion around vocabulary or concepts or misunderstandings related to response options that a researcher might overlook without cognitive interviewing. This provides additional assurance that the investigator and the respondent have a common understanding about the meaning of items and, thus, can enhance the validity of a scale.

A note of caution: Some researchers advocate having the target population give final approval of questionnaires. This is admirable and is likely to give the participants a greater sense of active participation in the research process; however, it is important to recognize the boundaries of respondents' expertise. They may know better than the investigator how they and people of similar backgrounds talk about various issues, and they are uniquely qualified to provide insights into their own understanding of a questionnaire's contents. Focus groups and cognitive interviewing can capitalize on the expertise that members of the target population have with respect to such matters. They are not, however, experts in the technical details of scale construction. It is unfair to expect nonexperts to understand the issues discussed in Chapter 5, for example, as they apply to item construction. As a case in point, a nonexpert might prefer an item that is worded in a pleasant, moderate way, whereas an experienced scale developer might recognize that the preferred wording would generate little variation in responses, thus rendering the item useless. My personal recommendation is this: Be sensitive to participants' concerns, help them feel actively involved in a variety of ways if that is appropriate, make full use of their expertise regarding linguistic or cultural aspects of item wording, and heed their observations regarding the meaning of items for the populations of interest, but reserve the right of final approval on item wording. We do not honor our research participants if we inadvertently create a situation in which their views, feelings, or opinions cannot be accurately gauged; we simply waste their time. The goal is to understand what ways of expressing concepts will be most clear to respondents while maintaining the integrity of those concepts. Doing so can benefit from understanding the special insights that the participants and investigators bring to the endeavor and capitalizing on both sets of strengths.

Decide on the Mode of Scale Administration

Researchers can collect data in a variety of ways (e.g., Dillman, 2007; Fowler, 2009), and they may choose to match modes of administration to the preferences of respondents. Accordingly, a team of investigators might consider using an interview rather than a printed questionnaire. It is important to recognize that a scale intended to be completed in print form may have substantially different properties when the items and responses are presented orally. For example, parents might be more reluctant to acknowledge high aspirations if they have to report them aloud to an interviewer rather than marking a response option. Generally, it is wise to restrict the mode of administering a new scale to the method used during scale development. Alternatively, researchers should systematically examine the impact of mode of administration on a scale. A generalizability study (see Chapter 3) may be used to determine the scale's generalizability across administration modes.

Consider the Scale in the Context of Other Measures or Procedures

What questions or research procedures will precede the scale itself? How will these questions affect responses to the scale? Nunnally (1978, pp. 627–677) refers to contextual factors such as response styles, fatigue, and motivation as *contingent variables.* He also points out that they can adversely affect research in three ways: (a) by reducing the reliability of scales; (b) by constituting reliable sources of variation other than the construct of interest, thus lowering validity; and (c) by altering the relationships among variables, making them appear, for example, to be more highly correlated than they actually are. As an example of how contingent variables might operate, consider *mood induction* and *cognitive sets* as they might apply to the marketing research example. The former might be an issue if, for example, the market researchers decided to include a depression or self-esteem scale in the same questionnaire as their aspirations scale. Scales tapping these (and other) constructs often contain items that express negative views of one's self. The Rosenberg Self-Esteem Scale (Rosenberg, 1965), for example, contains such items as "I feel I do not have much to be proud of" (as well as items expressing a positive self-appraisal). A researcher who was not sensitive to the potential effects of mood induction might select a series of exclusively self-critical items to accompany a newly developed scale. Reading statements that consistently express negative assessments of one's self may induce a dysphoric state that, in turn, can cause whatever follows to be perceived differently than it would otherwise (e.g., Kihlstrom, Eich, Sandbrand, & Tobias, 2000; Rholes, Riskind, & Lane, 1987). This might result in each of the three adverse effects noted by Nunnally (1978). That is, in the presence of affectively negative items, aspiration items might take on a somewhat different shade of meaning, thus lowering the proportion of variance in those items that is attributable to the intended latent variable. Or, in an extreme instance, some items from the aspiration scale might come to be influenced primarily by the induced mood state, rendering the scale multifactorial and lowering its validity as a measure of parental aspiration. Finally, to the extent that respondents' mood affected their responses to the aspiration items, scores for this scale might artificially correlate highly with other mood-related measures.

Cognitive sets are a more general example of the same phenomenon. That is, some frame of reference other than mood might be induced by focusing respondents' attention on a specific topic. For example, immediately preceding the aspiration scale with items concerning the respondents' income, the value of their home, and the amount they spend annually on various categories of consumer goods might create a mental set that temporarily alters their aspirations for their children. As a result, the responses to the scale might reflect an unintended transient state. As with mood, this cognitive set might adversely affect the reliability and/or validity of the scale by contaminating the extent to which it unambiguously reflects parental aspirations.

AFTER SCALE ADMINISTRATION

A quite different set of issues emerges after the scale has been used to address a substantive research question. The primary concerns at this point are analysis and interpretation of data generated by the instrument.

Analytic Issues

One issue in data analysis is the appropriateness of various techniques for variables with different scaling properties. The theoretical perspective and methods advocated most strongly in this book should result in scales that are amenable to a wide variety of data analytic methods. Although, strictly speaking, items using Likert or semantic differential response formats may be ordinal, a wealth of accumulated experience supports applying interval-based analytic methods to the scales they yield. However, the question of what methods are best suited to what types of data has been, and certainly will continue to be, hotly debated in the social sciences. Determining how different response options affect estimates of underlying variables is an active research area in its own right. Also, different audiences will have different expectations for how measures are treated. Whereas psychologists, for example, may be fairly sanguine in treating Likert scales as producing interval-level data, epidemiologists may not be. Perhaps the most practical approach is to monitor (and conform to) the prevailing sentiment with respect to this issue in one's area of interest.

Interpretation Issues

Assuming that the researcher has arrived at a suitable analytic plan for the data generated by a newly developed scale, the question of how to interpret the data remains. One point to keep in mind at this juncture is that the validity of a scale is not firmly established during scale development. Validation is a cumulative, ongoing process. Moreover, validity is really a characteristic of how a scale is used, not of the scale itself. A depression scale, for example, may be valid for assessing depression but not for assessing general negative affect.

Also, it is important to *think* about one's findings. Especially if the results appear strongly counterintuitive or countertheoretical, the researcher must consider the possibility that the scale is invalid in the context of that particular study (if not more broadly). It may be that the extent to which the validity of the scale generalizes across populations, settings, specific details of administration, or an assortment of other dimensions is limited. For example, the hypothetical parental aspiration measure might have been developed with a relatively affluent population in mind, and its validity for individuals whose resources are more limited may be unacceptable. Any conclusions based on a scale that has had limited use should consider the following: (a) how its present application differs from the context of its original validation, (b) the likelihood that those differences might limit the validity of the scale, and (c) the implications of those limitations for the present research.

Generalizability

The previous paragraph cautioned against generalizing across populations, settings, and other aspects of research. This issue warrants further emphasis. Reaching conclusions about group differences potentially confounds differences in the phenomenon being measured and differences in the performance of the instrument. If we can assume the latter to be trivial, then we can ascribe observed differences to group membership. In many situations (e.g., comparing off-task time in two groups of children randomly selected and allocated to groups), this will be the case. In others (e.g., comparisons across culturally distinct groups), we cannot assume identical instrument performance. Differential item functioning, discussed briefly in Chapter 7, is an active area of psychometric research. Although most researchers will not make it a focus of their own efforts, they should recognize the possibility of its presence and the limitations it may impose on their conclusions.

FINAL THOUGHTS

Measurement is a vital aspect of social and behavioral research. No matter how well designed and executed other aspects of a research endeavor may be, measurement can make or break a study. We assume that the variables of interest to us correspond to the assessment procedures we use. Often, the relationship of primary interest is between two or more unobservable variables, such as desire for a certain outcome and failure to consider alternative outcomes. We cannot directly measure desire or consideration, so we construct measures that we hope will capture them. These measures are, in a sense, quantitative metaphors for the underlying concepts. Only to the extent that those metaphors are apt (i.e., the instruments are valid) will the relationships we observe between measures reflect the relationship we wish to assess between unobservable constructs. Exquisite sampling, superb research design, and fastidious implementation of procedures will not change this fact. A researcher who does not understand the relationship between measures and the variables they represent, in a very literal sense, does not know what he or she is talking about. Viewed in this light, the efforts entailed in careful measurement are amply rewarded by their benefits.

References

Ader, D. N. (2007). Developing the Patient-Reported Outcomes Measurement Information System (PROMIS). *Medical Care, 45*(5, Suppl. 1), S1–S2.

Ajzen, I., & Fishbein, M. (1980). *Understanding attitudes and predicting behavior.* Englewood Cliffs, NJ: Prentice Hall.

Alder, K. (2002). *The measure of all things: The seven-year odyssey and hidden error that transformed the world.* New York: Free Press.

Allen, M. J., & Yen, W. M. (1979). *Introduction to measurement theory.* Monterey, CA: Brooks/Cole.

Anastasi, A. (1968). *Psychological testing* (3rd ed.). New York: Macmillan.

Asher, H. B. (1983). *Causal modeling* (2nd ed.). Sage university paper series on quantitative applications in the social sciences (Series No. 07-003). Beverly Hills, CA: Sage.

Barnette, W. L. (1976). *Readings in psychological tests and measurements* (3rd ed.). Baltimore: Williams & Wilkins.

Blalock, S. J., DeVellis, R. F., Brown, G. K., & Wallston, K. A. (1989). Validity of the Center for Epidemiological Studies Depression Scale in arthritis populations. *Arthritis and Rheumatism, 32,* 991–997.

Bohrnstedt, O. W. (1969). A quick method for determining the reliability and validity of multiple-item scales. *American Sociological Review, 34*, 542–548.

Bollen, K. A. (1989). *Structural equations with latent variables.* New York: Wiley.

Buchwald, J. Z. (2006). Discrepant measurements and experimental knowledge in the early modern era. *Archive for History of Exact Sciences, 60*, 565–649.

Campbell, D. T., & Fiske, D. W. (1959). Convergent and discriminant validation by the multitrait-multimethod matrix. *Psychological Bulletin, 56*, 81–105.

Carmines, E. G., & McIver, J. P. (1981). Analyzing models with unobserved variables: Analysis of covariance structures. In G. W. Bohrnstedt & B. F. Borgatta (Eds.), *Social measurement: Current issues* (pp. 65–115). Beverly Hills, CA: Sage.

Cattell, R. B. (1966). The screen test for the number of factors. *Multivariate Behavioral Research, 1*, 245–276.

Cohen, J. (1960). A coefficient of agreement for nominal scales. *Educational and Psychological Measurement, 20*(1), 37–46.

Cohen, P., Cohen, J., Teresi, J., Marchi, M., & Velez, C. N. (1990). Problems in the measurement of latent variables in structural equation causal models. *Applied Psychological Measurement, 14*, 183–196.

Comrey, A. L. (1973). *A first course in factor analysis.* New York: Academic Press.

Comrey, A. L. (1988). Factor analytic methods of scale development in personality and clinical psychology. *Journal of Consulting and Clinical Psychology, 56*, 754–761.

Converse, J. M., & Presser, S. (1986). *Survey questions: Handcrafting the standardized questionnaire.* Sage university paper series on quantitative applications in the social sciences (Series No. 07-063). Beverly Hills, CA: Sage.

Crocker, L., & Algina, J. (1986). *Introduction to classical and modern test theory.* New York: Holt, Rinehart & Winston.

Cronbach, L. J. (1951). Coefficient alpha and the internal structure of tests. *Psychometrika, 16,* 297–334.

Cronbach, L. J., Gleser, G. C., Nanda, H., & Rajaratnam, N. (1972). *Dependability of behavioral measurements: Theory of generalizability for scores and profiles.* New York: Wiley.

Cronbach, L. J., & Meehl, P. E. (1955). Construct validity in psychological tests. *Psychological Bulletin, 52,* 281–302.

Cureton, E. E. (1983). *Factor analysis: An applied approach.* Hillsdale, NJ: Lawrence Erlbaum.

Currey, S. S., Callahan, L. F., & DeVellis, R. F. (2002). *Five-item Rheumatology Attitudes Index (RAI): Disadvantages of a single positively worded item.* Unpublished paper, Thurston Arthritis Research Center, University of North Carolina at Chapel Hill.

Czaja, R., & Blair, J. (1996). *Designing surveys: A guide to decisions and procedures.* Thousand Oaks, CA: Pine Forge.

Dale, F., & Chall, J. E. (1948). A formula for predicting readability: Instructions. *Education Research Bulletin, 27,* 37–54.

Davey, Z. Z. T., & Hendrickson, A. (2010, May). *Classical versus IRT statistical test specifications for building test forms.* Paper presented at the annual meeting of the National Council of Measurement Education, Denver, Colorado.

De Boeck, P., & Wilson, M. (2004). *Explanatory item response models: A generalized linear and nonlinear approach.* New York: Springer.

DeVellis, R. F. (1996). A consumer's guide to finding, evaluating, and reporting on measurement instruments. *Arthritis Care and Research, 9,* 239–245.

DeVellis, R. F. (2005). Inter-rater reliability. In K. Kempf-Leonard (Ed.), *Encyclopedia of social measurement* (Vol. 2, pp. 317–322). San Diego: Elsevier.

DeVellis, R. F., & Callahan, L. F. (1993). A brief measure of helplessness: The helplessness subscale of the Rheumatology Attitudes Index. *Journal of Rheumatology, 20,* 866–869.

DeVellis, R. F., DeVellis, B. M., Blanchard, L. W., Klotz, M. L., Luchok, K., & Voyce, C. (1993). Development and validation of the Parent Health Locus of Control (PHLOC) scales. *Health Education Quarterly, 20,* 211–225.

DeVellis, R. F., DeVellis, B. M., Revicki, D. A., Lurie, S. J., Runyan, D. K., & Bristol, M. M. (1985). Development and validation of the child improvement locus of control (CILC) scales. *Journal of Social and Clinical Psychology, 3,* 307–324.

DeVellis, R. F., Holt, K., Renner, B. R., Blalock, S. J., Blanchard, L. W., Cook, H. L., et al. (1990). The relationship of social comparison to rheumatoid arthritis symptoms and affect. *Basic and Applied Social Psychology, 11,* 1–18.

DeWalt, D. A., Rothrock, N., Yount, S., & Stone, A. A. (2007). Evaluation of item candidates: The PROMIS qualitative item review. *Medical Care, 45*(5, Suppl. 1), S12–S21.

Dillman, D. A. (2007). *Mail and Internet surveys: The tailored design* (2nd ed., 2007 update). Hoboken, NJ: Wiley.

Duncan, O. D. (1984). *Notes on social measurement: Historical and critical*. New York: Russell Sage.

Embretson, S. E., & Hershberger, S. L. (1999). Summary and future of psychometric models in testing. In S. E. Embretson & S. L. Hershberger (Eds.), *The new rules of measurement* (pp. 243–254). Mahwah, NJ: Lawrence Erlbaum.

Embretson, S. E., & Reise, S. P. (2010). *Item response theory* (2nd ed.). New York: Routledge Academic.

Fan, X. (1998). Item response theory and classical test theory: An empirical comparison of their item/person statistics. *Educational and Psychological Measurement, 58*, 357–381.

Festinger, L. (1954). A theory of social comparison processes. *Human Relations, 7*, 117–140.

Fink, A. (1995). *The survey kit*. Thousand Oaks, CA: Sage.

Fowler, F. J., Jr. (2009). *Survey research methods* (4th ed.). Thousand Oaks, CA: Sage.

Fry, E. (1977). Fry's readability graph: Clarifications, validity, and extension to level 17. *Journal of Reading, 21*, 249.

Ghiselli, B. E., Campbell, J. P., & Zedeck, S. (1981). *Measurement theory for the behavioral sciences*. San Francisco: Freeman.

Gorsuch, R. L. (1983). *Factor analysis*. Hillsdale, NJ: Lawrence Erlbaum.

Hambleton, R. K., Swaminathan, H., & Rogers, H. J. (1991). *Fundamentals of item response theory*. Newbury Park, CA: Sage.

Harman, H. H. (1976). *Modern factor analysis*. Chicago: University of Chicago Press.

Hathaway, S. R., & McKinley, J. C. (1967). *Minnesota Multiphasic Personality Inventory: Manual for administration and scoring*. New York: Psychological Corporation.

Hathaway, S. R., & Meehl, P. E. (1951). *An atlas for the clinical use of the MMPI*. Minneapolis: University of Minnesota Press.

Hayton, J. C., Allen, D. G., & Scarpello, V. (2004). Factor retention decisions in exploratory factor analysis: A tutorial on parallel analysis. *Organizational Research Methods, 7*(2), 191–205.

Idler, E. L., & Benyamini, Y. (1997). Self-rated health and mortality: A review of twenty-seven community studies. *Journal of Health and Social Behavior, 38*, 21–37.

Jöreskog, K. G. (1971). Simultaneous factor analysis in several populations. *Psychometrika, 36*, 109–134.

Kaiser, H. F. (1960). The application of electronic computers to factor analysis. *Educational and Psychological Measurement, 20*, 141–151.

Keefe, F. J. (2000). Self-report of pain: Issues and opportunities. In A. Stone, J. S. Turkkan, C. A. Bachrach, J. B. Jobe, H. S. Kurtzman, & V. S. Cain (Eds.), *The science of self-report: Implications for research and practice* (pp. 317–337). Mahwah, NJ: Lawrence Erlbaum.

Kelly, J. R., & McGrath, J. B. (1988). *On time and method*. Newbury Park, CA: Sage.

Kihlstrom, J. F., Eich, E., Sandbrand, D., & Tobias, B. A. (2000). Emotion and memory: Implications for self-report. In A. Stone, J. S. Turkkan, C. A. Bachrach,

J. B. Jobe, H. S. Kurtzman, & V. S. Cain (Eds.), *The science of self-report: Implications for research and practice* (pp. 81–103). Mahwah, NJ: Lawrence Erlbaum.

Kirk, R. E. (1995). *Experimental design: Procedures for the behavioral sciences* (3rd ed.). San Francisco: Brooks/Cole.

Krueger, R. A., & Casey, M. A. (2000). *Focus groups: A practical guide for applied research*. Thousand Oaks, CA: Sage.

Levenson, H. (1973). Multidimensional locus of control in psychiatric patients. *Journal of Consulting and Clinical Psychology, 41*, 397–404.

Lipsey, M. W. (1990). *Design sensitivity: Statistical power for experimental research*. Newbury Park, CA: Sage.

Loehlin, J. C. (1998). *Latent variable models: An introduction to factor, path, and structural analysis*. Mahwah, NJ: Lawrence Erlbaum.

Long, J. S. (1983). *Confirmatory factor analysis*. Sage university paper series on quantitative applications in the social sciences (Series No. 07-033). Beverly Hills, CA: Sage.

Lord, F. M., & Novick, M. R. (2008). *Statistical theories of mental test scores*. Charlotte, NC: Information Age.

MacCallum, R. C., Widaman, K. F., Zhang, S., & Hong, S. (1999). Sample size in factor analysis. *Psychological Methods, 4*, 84–99.

Mayer, J. M. (1978). Assessment of depression. In M. P. McReynolds (Ed.), *Advances in psychological assessment* (Vol. 4, pp. 358–425). San Francisco: Jossey-Bass.

McDonald, R. P. (1984). *Factor analysis and related methods*. Hillsdale, NJ: Lawrence Erlbaum.

Messick, S. (1995). Validity of psychological assessment: Validation of inferences from persons' responses and performances as scientific inquiry into score meaning. *American Psychologist, 50*, 741–749.

Mitchell, S. K. (1979). Interobserver agreement, reliability, and generalizability of data collected in observational studies. *Psychological Bulletin, 86*, 376–390.

Mlodinow, L. (2008). *The drunkard's walk: How randomness rules our lives*. New York: Pantheon.

Murphy, L. L., Spies, R. A., & Plake, B. S. (2006). *Tests in print*. Lincoln: University of Nebraska Press.

Myers, J. L. (1979). *Fundamentals of experimental design* (3rd ed.). Boston: Allyn & Bacon.

Namboodiri, K. (1984). *Matrix algebra: An introduction*. Sage university paper series on quantitative applications in the social sciences (Series No. 07-028). Beverly Hills, CA: Sage.

Narens, L., & Luce, R. D. (1986). Measurement: The theory of numerical assignments. *Psychological Bulletin, 99*, 166–180.

Nering, M. L., & Ostini, R. (2010). *Handbook of polytomous item response theory models*. New York: Routledge.

Nunnally, J. C. (1978). *Psychometric theory* (2nd ed.). New York: McGraw-Hill.

Nunnally, J. C., & Bernstein, I. H. (1994). *Psychometric theory* (3rd ed.). New York: McGraw-Hill.

Osgood, C. E., & Tannenbaum, P. H. (1955). The principle of congruence in the prediction of attitude change. *Psychological Bulletin, 62*, 42–55.

PROMIS. (2010). *Patient-Reported Outcomes Measurement Information System: Dynamic tools to measure health outcomes from the patient perspective.* http://www.nihpromis.org/default.aspx

Radloff, L. (1977). The CES-D scale: A self-report depression scale for research in the general population. *Applied Psychological Measurement, 1*, 385–401.

Rasch, G. (1960). *Probabilistic models for some intelligence and attainment tests.* Chicago: MESA.

Rholes, W. S., Riskind, J. H., & Lane, J. W. (1987). Emotional states and memory biases: Effects of cognitive priming and mood. *Journal of Personality and Social Psychology, 52*, 91–99.

Robinson, J. P., Shaver, P. R., & Wrightsman, L. S. (1991). *Measures of personality and social psychological attitudes*. San Diego: Academic Press.

Rosenberg, M. (1965). *Society and the adolescent self-image*. Princeton, NJ: Princeton University Press.

Rotter, J. B. (1966). Generalized expectancies for internal vs. external control of reinforcement. *Psychological Monographs, 80*(1, Whole No. 609).

Samejima, F. (1969). Estimation of latent ability using a response pattern of graded scores. *Psychometric Monograph* (Suppl. 17).

Saucier, G., & Goldberg, L. R. (1996). The language of personality: Lexical perspectives on the five-factor model. In J. S. Wiggins (Ed.), *The five-factor model of personality* (pp. 21–50). New York: Guilford.

Shrout, P. E., & Fleiss, J. L. (1979). Intraclass correlations: Uses in assessing rater reliability. *Psychological Bulletin, 86*, 420–428.

Sijtsma, K. (2009). On the use, the misuse, and the very limited usefulness of Cronbach's alpha. *Psychometrika, 74*, 107–120.

Silvestro-Tipay, J. L. (2009). Item response theory and classical test theory: An empirical comparison of item/person statistics in a biological science test. *International Journal of Educational and Psychological Assessment, 1*, 19–31.

Smith, P. H., Earp, J. A., & DeVellis, R. F. (1995). Measuring battering: Development of the Women's Experiences with Battering (WEB) scale. *Women's Health: Research on Gender, Behavior, and Policy, 1*, 273–288.

Spielberger, C. D., Gorsuch, R. L., & Lushene, R. E. (1970). *State-trait anxiety inventory (STAI) test manual for form X.* Palo Alto, CA: Consulting Psychologists Press.

Spies, R. A., Carlson, J. F., & Geisinger, K. F. (2010). *The eighteenth mental measurements yearbook.* Lincoln: University of Nebraska Press.

Stage, C. (2003). *Classical test theory or item response theory: The Swedish experience* (EM No. 42). Umeå, Sweden: Umeå Universitet Department of Educational Measurement. Available online at http://www8.umu.se/edmeas/publikationer/index_eng.html

Sterba, K. R., DeVellis, R. F., Lewis, M. A., Baucom, D. H., Jordan, J. M., & DeVellis, B. M. (2007). Developing and testing a measure of dyadic efficacy for married women with rheumatoid arthritis and their spouses. *Arthritis & Rheumatism (Arthritis Care & Research), 57*(2), 294–302.

Strahan, R., & Gerbasi, K. (1972). Short, homogenous version of the Marlowe-Crowne Social Desirability Scale. *Journal of Clinical Psychology, 28*, 191–193.

Thissen, D., Chen, W.-H., & Bock, R. D. (2003). Multilog (Version 7.0) [Computer software]. Lincolnwood, IL: Scientific Software International.

Tinsley, H. E. A., & Tinsley, D. J. (1987). Uses of factor analysis in counseling psychology research. *Journal of Counseling Psychology, 34*, 414–424.

Van der Linden, W. J., & Glas, C. A. W. (2000). *Computerized adaptive testing: Theory and practice*. St. Paul, MN: Assessment Systems.

Wallston, K. A., Stein, M. J., & Smith, C. A. (1994). Form C of the MHLC Scales: A condition-specific measure of locus of control. *Journal of Personality Assessment, 63*, 534–553.

Wallston, K. A., Wallston, B. S., & DeVellis, R. (1978). Development and validation of the multidimensional health locus of control (MHLC) scales. *Health Education Monographs, 6*, 161–170.

Weisberg, H., Krosnick, J. A., & Bowen, B. D. (1996). *An introduction to survey research, polling, and data analysis*. Thousand Oaks, CA: Sage.

Willis, G. B. (2005). *Cognitive interviewing: A tool for improving questionnaire design*. Thousand Oaks, CA: Sage.

Wright, B. D. (1999). Fundamental measurement for psychology. In S. E. Embretson & S. L. Hershberger (Eds.), *The new rules of measurement* (pp. 65–104). Mahwah, NJ: Lawrence Erlbaum.

Yu, C. H. (2005). Test-retest reliability. In K. Kempf-Leonard (Ed.), *Encyclopedia of social measurement* (Vol. 3, pp. 777–784). San Diego: Elsevier.

Zickar, M. J., & Broadfoot, A. A. (2008). The partial revival of a dead horse? Comparing classical test theory and item response theory. In C. E. Lance & R. J. Vandenberg (Eds.), *Statistical and methodological myths and urban legends*. New York: Routledge Academic.

Zorzi, M., Priftis, K., & Umilitá, C. (2002, May). Brain damage: Neglect disrupts the mental number line. *Nature, 417*, 138–139.

Zuckerman, M. (1983). The distinction between trait and state scales is not arbitrary: Comment on Allen and Potkay's "On the arbitrary distinction between traits and states." *Journal of Personality and Social Psychology, 44*, 1083–1086.

Zwick, W. R., & Velicer, W. F. (1986). Comparison of five rules for determining the number of components to retain. *Psychological Bulletin, 99*, 432–442.

Index